SCIENCE ACTIVITIES
FOR CHILDREN

Willard J. Jacobson

Teachers College
Columbia University

Abby B. Bergman

The Ralph S. Maugham School
Tenafly, New Jersey

PRENTICE-HALL, INC., ENGLEWOOD CLIFFS, NEW JERSEY 07632

Library of Congress Cataloging in Publication Data

Jacobson, Willard J.
    Science activities for children.

    1. Science—Experiments.   I. Bergman,
Abby Barry. II. Title.
Q164.J3   507'.8   82-7606
ISBN 0-13-794594-9   AACR2

TO THE CHILDREN WHO HAVE TAUGHT US

Interior design and editorial supervision: Serena Hoffman
Cover design: Wanda Lubelska
Manufacturing buyer: Ron Chapman

© 1983 by Prentice-Hall, Inc., Englewood Cliffs, N.J. 07632

Printed in the United States of America

10 9 8 7 6 5 4

ISBN 0-13-794594-9

PRENTICE-HALL INTERNATIONAL, INC., *London*
PRENTICE-HALL OF AUSTRALIA PTY LIMITED, *Sydney*
EDITORA PRENTICE-HALL DO BRAZIL, LTDA, *Rio de Janeiro*
PRENTICE-HALL CANADA INC., *Toronto*
PRENTICE-HALL OF INDIA PRIVATE LIMITED, *New Delhi*
PRENTICE-HALL OF JAPAN, INC., *Tokyo*
PRENTICE-HALL OF SOUTHEAST ASIA PTE. LTD., *Singapore*
WHITEHALL BOOKS LIMITED, *Wellington, New Zealand*

Contents

[1]*P* means *Primary,* or an activity for kindergarten through second grade; *I* means *Intermediate,* or an activity for grades three and up; *P, I* means an activity suitable for both primary and intermediate levels.

4 STUDYING OURSELVES 97

5 **THE EARTH   119**

6 **MAGNETISM AND ELECTRICITY   135**

7 MACHINES   161

8 EXPLORING THE UNIVERSE: SUN, MOON, AND STARS   181

Preface

Science Activities for Children will be of value not only to teachers, but also to anyone who works and plays with children—parents, grandparents, teachers, people who work in day care centers, scout leaders, summer camp personnel, and anyone else interested in helping one child or many have rewarding experiences in science. We hope you will enjoy these experiences. We know the children will.

Experiences in science are very important for the intellectual and physical development of children. Children have a chance to manipulate, handle, and experiment with plants, magnets, wire, mirrors, and many other concrete materials. While words and abstract ideas are very important to children's development, it is critical that they also have firsthand experiences with the concrete materials and activities that the words refer to.

In science we can begin to have experiences that are the foundations of logical thought. Most of the activities in this book lead children to investigate questions and problems. From these experiences, they learn to inquire into the nature of the world they live in. Science activities, which lead children into firsthand experiences with concrete materials and which lead them to develop inquiry skills, are considered to be critical for the optimum intellectual development of children.

Science can be fun. The science activities described in this book have been fun for hundreds of youngsters. We hope your children, too, can have such fun. But children can also gain a better understanding from these activities of some of the most important ideas in science. Such ideas are discussed in each investigation under "Background Information." These are ideas that children can carry with them to other experiences in life.

Almost all of the science activities in this book can be used in the school and the home. However, there are many other situations in which science activities can be undertaken. Try science in such situations as:

When a child says "There is just nothing to do."
When it's nice to be out-of-doors.
When it is raining and you wonder what you can do.

At night when the stars are out.
In the daytime when the sun is in the sky.
On a cloudy day.
In the kitchen.
At summer camp.
With scouting groups.
With small groups of children.
With large groups of children.
As community activities.
In the garden.
On the school grounds.
When you are faced with three (or one, or two, or four, or five) bored children.
When the idea you have been talking about is not clear, and handling some concrete materials will help.

You can find science activities in this book that will be useful in such situations. Children, of course, should have the fun of doing these activities, and they can begin to gain concepts of the "big ideas" discussed under "Background Information." You as an adult can also participate. You, too, can enjoy these activities, and the children can gain from your participation.

A wide variety of science activities is included in this book. Each activity has a title and an investigation. After the investigation, there are letters designating the grade levels these activities are recommended for. The recommended levels for activities are also indicated in the table of contents. P indicates *primary*, and that activity can be used with very young children. In schools, activities designated P can be used in the kindergarten and through the second grade. I indicates *intermediate*, and that activity can be used with children aged eight and older. In schools, activities designated I can be used in grades three and up. Many activities are designated both P and I, meaning that these activities can be used at both levels.

The metric system is used throughout much of the book. The metric system is the system of measurement used in most of the sciences, and children should have experience using it. Of course, for most purposes it is also much simpler and easier to learn than the English System. Therefore, to help children become acquainted with the metric system, the equivalents in the English System are also given throughout the book.

We believe that the sections entitled "Going Further" are of special importance. Here are suggestions for enrichment—activities for highly motivated children and children who have special aptitudes and interests in science. Certainly, all children should be helped to achieve optimal intellectual development. The science activities described in this book can help nurture the aptitudes and capacities of many who have a special bent for science.

Most of the materials and equipment needed for these science activities can be obtained in the home, school, or local hardware and grocery store.

However, we suggest that you check the list of materials needed before you undertake a science activity. "For want of a paper clip, the ardor cooled."

Of course, the amount of materials needed will depend upon the number of children. However, even when you work with groups of children, each child should have a chance to handle the equipment and materials. These are "hands on" activities.

There is no special sequence to the activities. You can start at the beginning and work from cover to cover. Or you can choose a chapter and complete those activities. Or you can choose almost any activity in the book and start with it. But do start! Many children have already gone too long without this kind of experience in science.

ACKNOWLEDGMENTS

"To the children who have taught us."

We are indebted to the hundreds of children with whom we have worked. Many of the science activities in this book are new to the science education literature, and some of the traditional activities have new and different dimensions. Many of these new ideas are the result of insights gained through work with children. They have taught us.

We are also indebted to the thousands of dedicated teachers with whom we have worked. Their idealism and dedication have inspired us. The children they teach are fortunate to have learned and grown with them.

We have learned a great deal from fellow professionals in elementary science education. The meetings of the Council for Elementary Science International (CESI) have always been rewarding. We have gained from our participation in the Science Curriculum Improvement Study (SCIS) and the American Association for the Advancement of Science's *Science—A Process Approach* (SAPA). The discussions with scientists and science educators at the New York Academy of Sciences have always been stimulating. We have always benefitted from the sharp, critical discussions in the Department of Science Education at Teachers College, Columbia University.

Many have helped and stimulated us, but we would be derelict if we did not specifically mention a few. Mitchell Batoff, Bonnie Brownstein, Albert Carr, Morsley Giddings, Elizabeth Meng, Lester Mills, Bernadette Morton, Catherine Namuddu, Herma Perkins, Dorothy Pottinger, Harold Tannenbaum, Robert Walrich, Wendy Wetterer, and John Youngpeter have shared our delight with elementary science and have contributed many ideas in more ways than they, or we, may be aware. We are indebted to Thomas Caynon for help in the preparation of the manuscript.

Our wives, Carol and Rose, have been supportive, helpful, and inspirational. Without them we could not have written this book, and we are deeply grateful.

WILLARD J. JACOBSON
ABBY BARRY BERGMAN

Plants and Animals

INTRODUCTION

Few experiences in a child's life can be as fulfilling and rewarding as caring for and watching the growth and development of living things. Whether it be training a dog at home, setting up and maintaining a terrarium for exotic tropical plants, or simply growing a marigold plant in an empty milk carton, there is something unique about these activities that has a special appeal for youngsters.

The study and care of plants can help children learn a great deal about living organisms. All youngsters should have a chance to grow plants, care for them, study their needs, and observe the changes that occur in them. Those individuals who guide children as they explore the plant world should be careful to convey positive attitudes that promote the conservation of living things and their natural environments. A basic premise in obtaining plants for study is to never take anything (plant or animal) from a natural environment which would in any way result in a change of that environment. Of course, our judgment ought to be ruled with reason. No appreciable environmental change would result from each child in a group's taking a dandelion from a field that contains thousands; but plucking a single spray of wild flowers from a roadside where no others like it are seen may well have a detrimental impact upon the environment.

Working with and caring for animals makes it possible for children to study firsthand some of the major concepts of animal life as we know it. Providing a suitable habitat for an animal, watching it in action, learning its ways, and being responsible for its general well-being are all important tasks for children to assume. Along with the joys and pleasures of keeping animals come important responsibilities that children are never too young to develop. The animal's environment should be carefully planned before the animal is acquired. Children can be encouraged to gather materials that, when placed together, replicate as much as possible the animal's natural surroundings. Sufficient room for free movement of the animal should always be provided. Procedures for its care, feeding, cleaning, and boarding during vacation periods should all be arranged with the group of children assuming responsibility for the animal. Manuals and guides for the care of the particular type of animal are most helpful. As the children observe and study the animal's habits, it should never be subjected to any treatment or deprivation that might result in pain or injury.

Caring for animals is an interest that is usually developed in childhood, but it is an interest that can provide lifelong fulfillment. Many adults can be seen tending a collection of tropical fish with patience, love, and enthusiasm. With positive leadership and the development of respect for all of nature, children will surely acquire the beginnings of the enjoyment of plants and animals for the rest of their lives. The activities that follow may point the way!

PLANTS IN THE NEIGHBORHOOD

Investigation: What kinds of plants can be found in the local community?   (P, I)

Background Information: A great variety of plants can be found in any school community. Differences in size, color, texture, and environment are relatively easy to discern. Children can learn to identify common plants in the school neighborhood as well as major plant categories—for example, flat, leafy (broadleaf) plants and plants with needlelike leaves (conifers).

Materials: Spoons or small shovels for digging
Magnifiers
Plant or tree guides

Procedure: 1. Take the group on a walk through the community and point out different plants along the way. Ask the children if they know the names of any of the plants or trees they see.

2. Select a spot where a variety of different plants of varying heights can be observed.

3. Assign groups of children to look for plants which do not rise above their ankles, those which rise above their ankles but are not taller than they are, and those which are taller than they are.

4. Distribute materials for observing and removing, where appropriate, a whole plant. (Teacher judgment should be exercised so as to promote positive conservation attitudes.)

5. Ask the children to locate and list (or draw) plants that demonstrate the following extreme characteristics:
Largest and smallest
Most colorful and least colorful
Largest leaves and smallest leaves
Largest seeds and smallest seeds

6. Point out trees which have flat (broadleaf) leaves and those which have needlelike or scaly leaves (conifers).

7. Demonstrate the difference between a shrub and a tree. (Shrubs have several small trunks, while trees have a single large one.)

8. Try to distinguish between plants which were put there by people and plants which were not.

9. List the various environments in which the plants were found—for instance, vacant lots, sidewalk cracks, around telephone poles, or in gardens and parks.

Going Further: Have plant or tree guides available for children to use in identifying the specific names of plants they found during their walk. Some common plants can be pressed or ironed in a "sandwich" between two pieces of waxed paper. (Use a sheet of plain typing paper between the iron and the piece of waxed paper to keep the iron clean.) Interrelationships between plants can also be explored—for example, vines can often be seen growing around tree trunks. Some children can make booklets that include labeled drawings or pressed samples of plants found in the local community.

FINDING SEEDS

Investigation: Where do seeds come from?   (P)

Background Information: Seeds are found within the fruits of plants and vary from one another in several ways. Children can explore the sources of seeds available to them as well as the great variety of seeds easily uncovered. Physical differences among seeds, that is, differences in color, size, shape, and texture, are relatively easy to discern. It may also interest the children to note the number of seeds found in each of several apples, oranges, pears, or other fruit.

Materials: A variety of fruit: apples, oranges, pears, tomatoes, peaches
Plastic knives with serrated blades
Magnifiers

Procedure:
1. Discuss seeds and where they come from. What do seeds do if planted in soil?
2. Cut open an apple and remove the seeds. How many are there? How are they alike and different? Cut open a few apples. Do they all have the same number of seeds in them?
3. Remove the seeds from a variety of fruits. Ask the children to compare the color, size, shape, and texture of the various seeds. Can they match the seeds with the fruits they come from? Which fruits have the most seeds? Which fruits have only one seed?
4. Take a walk with the children and ask them to find seeds in the local community. Maple seeds are common during certain seasons, as are acorns (seeds from oak trees). The white cottony head of a dandelion contains thousands of seeds.
5. Encourage the children to make a collection of as many different types of seeds as they can find. Ask the children to sort the seeds according to a variety of criteria. For example, seeds can be sorted according to color, size, shape, origin, or other characteristics which the children specify. Ask the youngsters to talk about the reasons for their groupings.

Going Further: Almost all seeds can be germinated in the classroom. It is usually a good idea to soak the seeds overnight before attempting to plant them. Most seeds can be planted in potting soil or vermiculite. Others, such as avocado seeds, are best started in a jar of water. Which seeds produce little leaves or shoots soonest after planting?

Many crafts projects use seeds for decorative purposes. Children can use a variety of seeds to make pictures. They should first plan their pictures by arranging the seeds on paper or by outlining with a pencil areas which will be filled in with different kinds of seeds. Once these pictures have been planned, white glue may be placed in the areas upon which the seeds are to be glued, and then the seeds dropped into position. Seed necklaces and other ornaments are also interesting and fun to make.

INSIDE SEEDS

Investigation: What is inside of a seed?   (P, I)

Background Information: Plant seeds contain the necessary ingredients for the initial growth of a new plant. A baby plant (embryo) as well as food for the sustenance of this baby plant are contained in the seed. Sufficient nutrients are stored in the seed to maintain the plant until it can make food on its own. Seeds vary in the amount of stored food they contain and in their relative hardness; thus certain seed types are more appropriate for classroom study than others. Lima beans, mung beans, kidney beans, pinto beans, and other large beans are best for classroom use.

Materials: A box of lima beans
Plastic tumblers
Magnifiers
Blotter paper or heavy construction paper

Procedure: 1. Soak lima beans in a jar or tumbler overnight. (Plan on using three or four beans per child.)
2. The next day, distribute the soaked beans and magnifiers, and ask the children to open them carefully. What do they see? The tiny plants (embryos) should appear as very small leaves. The rest of the seed is stored food.
3. Ask the children to compare soaked beans with dry beans. How do they differ?
4. Soak other types of beans and ask the children to compare the appearance of the embryos and the amount of stored food.
5. Place some beans between the inside walls of the plastic tumblers and blotter paper which is pressed up along the sides of the tumbler to hold the beans in place (see Figure 1-1). Fill the tumber with water. Observe for several days. The stored food will eventually shrivel up and gradually disappear as the embryo uses it for its "start" as a plant.
6. As the plant grows, its main parts (roots, stem, leaves) are easy to see.

Going Further: Experiment with different types of seeds. Which grow the fastest? After the main parts of the plant have appeared, try different planting media (garden soil, sand, vermiculite) to see which results in the fastest growth of the new plant. Plants may be started in the individual compartments of egg cartons for ease of comparison. Does the depth at which the young plant is placed in the soil or sand affect its rate of growth? Why?

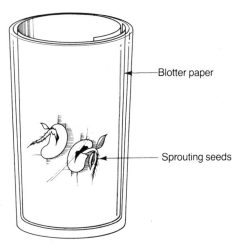

—Blotter paper

—Sprouting seeds

FIGURE 1-1. Root and leaf development can be observed as growing seeds are pressed up along the sides of a clear plastic tumbler.

EXPLORING STEMS

Investigation: What do stems do for a plant? (I)

Background Information: Stems of plants carry water and essential nutrients from the soil to other parts of the plant. The leaves in particular need nutrients, as they are the sites of food production. Little tubes, or vessels, within the stems transport the water and nutrients through the plant. In this activity children will be able to actually see these tiny tubes.

Materials: A few stalks of celery
Two plastic tumblers
Red ink or food coloring
A knife

Procedure: 1. Cut off the end (very wide, white part) of a large celery stalk and let it freshen in a glass of water for about 30 minutes.
2. Slice the celery stalk through the middle, the long way, halfway up its length. Tie, rubber band, or tape the stem where it is whole to prevent further splitting.
3. Place red ink or food coloring in one tumbler and water in the other. Place one half of the stalk into the tumbler of ink and the other half into the tumbler of water (see Figure 1-2).
4. Ask the children to predict what will happen.
5. Let the celery stalk stand in the containers for a few hours, and observe the leaves and stem at regular intervals.
6. After a few hours, remove the celery stalk from the tumblers and slice the stalk into several short lengths. Can the tiny transport tubes be seen? Are the tubes on one side of the stalk red in color? Why is this so? What conclusions can be drawn from this observation?
7. Look at the leaves on the celery stalk. Have any of them changed color? If so, why did this happen? What does this observation indicate about the internal structure of the leaves?

Going Further: Different shades of ink or food coloring may be used with individual stalks of celery to produce the effects of transfer within the stems. Put the cut ends of several types of stems or shoots into colored water or ink. Which ones seem to carry the ink through the plant the best? Cut short sections of the stems. In some plants, the color may only rise through a small portion of the stem.

Try white carnations in different glasses of colored water or ink. The results are usually dramatic!

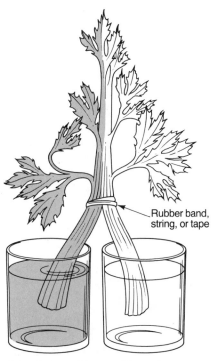

Rubber band, string, or tape

Colored water or ink   Clear water

FIGURE 1-2. The colored water or ink passes through the tiny tubes in the stem of the celery.

GROWING PLANTS FROM CUTTINGS

Investigation: How can plants be grown from cuttings? (P, I)

Background Information: A cutting is a small part of a plant that is removed and rooted in water. What could be simpler! Cuttings (sometimes called slips) are usually 5 to 10 centimeters (2 to 4 inches) long and are taken from shoots or new growth on the parent plant. Starting plants from cuttings offers the unique advantage of allowing the children to observe root development over a period of time. Among the list of plants that can be started from cuttings are philodendron, wandering Jew, geraniums, coleus, petunias, English ivy, wax begonia, pussy willow, and pachysandra.

Materials: A few cuttings from the growth shoots of one of the plants mentioned above
A glass or vase

Procedure:
1. Look for a short, stocky growth shoot on the plant from which the cutting is to be taken. Count 3 to 5 nodes (joints where leaves join the stem) from the tip of the shoot, and make a diagonal cut with a knife or razor just below a node (see Figure 1-3).
2. Remove the lowermost leaves, but allow a few of the younger leaves to remain.
3. Place the cutting in a vase, glass, or jar. To retain moisture, the cutting can be covered with plastic, but the plastic should be supported in such a way that it *does not touch the leaves.*
4. The cutting should be kept out of direct sunlight for the first few days. Replace water as it evaporates.
5. After a few weeks (if roots have grown), the cutting should be planted in a pot. A mixture of two-thirds potting soil and one-third vermiculite or sand makes an excellent potting medium. Do not attempt to plant the cutting if sufficient root growth is not evident.

Going Further: Cuttings from different plants may be started at the same time and root growth compared. Which plants develop roots first?

Plant hormones which promote root development are available in most plant or garden shops. Children may test the effects of these hormones by starting identical cuttings at the same time—one which has been treated with the hormone and one without such treatment. What happens?

Some cuttings develop better if started in damp sand or vermiculite. Which plants fare better if the cuttings are placed in one of these mediums?

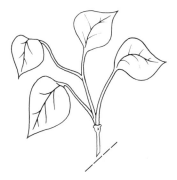

a. Cut stem just below a node

b. Remove all but 2 or 3 leaves

c. Insert cut stem into jar of water.

d. Cover with plastic.

FIGURE 1-3. Steps in starting plants with cuttings.

A PLANT FROM A POTATO

Investigation: How can a plant grow from a potato?   (P, I)

Background Information: A potato is a fleshy underground stem called a *tuber*. This stem contains stored food, mainly starch. The *eyes* of the potato are actually buds from which new growth begins. A sweet potato is an underground *root*. It consists of stored starch and sugar. Children are often amazed (as are many adults for that matter) when they witness the dense foliage which often results from the simple placement of a potato or sweet potato in a jar of water. It is the stored food contained in the potato or sweet potato which provides the nourishment for the growth of leaves. When potatoes are planted in a garden, cut-up sections of an old potato with a few eyes in each section are used to start the new plant.

Materials: A potato or sweet potato
A jar
Toothpicks

Procedure: 1. Place a potato or sweet potato in a jar of water and insert tooth-picks into the potato so that half of it is immersed in water when the toothpicks rest on the top of the jar. If a sweet potato is used, place the root or narrow end down into the water. If the potato is large enough, it may not require toothpicks for suspension, but will simply rest on the rim of the jar (see Figure 1-4).

FIGURE 1-4. When partially immersed in water, a potato or sweet potato will begin to grow.

2. Keep the jar and potato in a dark place for a few days. After some roots have formed, place on a windowsill or other light spot.
3. After roots and leaves have grown, the potato may be planted in soil.

Going Further: Other materials may be used to start a potato plant in. Sawdust, pebbles, sand, or vermiculite would be interesting to experiment with. Make sure that the material used is kept moist.

A sweet potato may be planted "upside-down." Roots will grow from one end of the sweet potato, while leaves will grow from the other end. An upside-down sweet potato will be slow to develop, but eventually the leaves and roots will reverse themselves—the roots will grow down into the water and the leaves will push up out of the jar into the air.

Carrot, beet, turnip, or parsnip tops may also be grown. Remove the old leaves and slice off the top portion of any of these vegetables and place in a shallow dish or jar top of water. Even the leaves of a pineapple top will continue to grow if placed in a dish of water.

THE NEEDS OF GREEN PLANTS

Investigation: What do green plants need to live and grow?   (P, I)

Background Information: In this investigation, children will determine the needs of green plants by designing a series of controlled experiments. By isolating variables (for example, the presence of water and sunlight) and periodically inspecting the condition of the plants, the children will be able to draw some conclusions about the requirements of the plants.

Materials: A small bag of potting soil or garden soil
Eight small paper cups
A packet of seeds (mung or kidney beans, or zinnia, sunflower, or marigold seeds)
Two lightproof cardboard boxes (shoe boxes will do)

Procedure:

1. Punch a small hole in the bottom of each of the paper cups and fill them with soil.
2. Place two or three seeds (or beans) under a centimeter (half inch) of soil in each of the cups.
3. Add water to the cups and place them on a window sill, watering them every few days until seedlings (tiny plants) appear.
4. Discuss the requirements of plant and children. List their suggestions. (Typical responses will include sunlight, water, air, soil, and so on.)
5. Ask the children if they can devise a way to test whether the plants actually need these conditions.
6. Elicit the need to conduct a controlled experiment in which certain variables (sun, water, air) are isolated and removed from a plant's environment. Use one plant as a "control," that is, a plant which is not subjected to any deprivation.
7. At first, it will be easiest to work with just two variables, for example, sun and water.
8. Label each of the test plants according to the conditions that they will be exposed to. For example:

Plant	Environment
Plant A (the control)	Sun, water
Plant B	Sun, but no water
Plant C	Water, but no sun
Plant D	No water, no sun

9. Place the plant which is to receive water but no sun in a lightproof cardboard box, but remove it from the box every other day to water it.

10. Place the plant which is to receive no water and no sun in a lightproof cardboard box, and do not water it again until the investigation is concluded.

11. Check the plants according to the record form that follows, and have the children keep records using the symbols indicated. This form may be duplicated on a spirit or mimeo master so that each child can keep his or her own records.

+ = Alive and growing
− = Dying, looks unhealthy
0 = Dead

RECORD FORM

	Plant A (the control)	Plant B (sun, no water)	Plant C (water, no sun)	Plant D (no sun, no water)
1st Day				
2nd Day				
5th Day				
8th Day				
14th Day				

12. What conclusions can the children draw from the results of their investigation?

Going Further: Some children might want to isolate other variables, such as the environmental temperature or the presence of air. Others might want to experiment with different varieties of plants to see which ones can withstand the longest periods of environmental deprivation. More complex investigations can be undertaken by working with three variables at the same time. How many plants would be needed to conduct a controlled experiment if water, sun, and air were the variables? The effects of different types of soil may also be explored. How do similar plants thrive in potting soil, sand, vermiculite, or garden soil?

ADOPT A TREE

Investigation: How can a tree be "adopted"?   (P, I)

Background Information: A child or group of children may enjoy the experience of "adopting" a particular tree. An accessible tree which may be readily observed without danger or obstruction should be selected. Through guided observations, children can focus on such characteristics as the tree's height, color, leaves, buds, bark, branching patterns, fruit, and the like. On closer examination, ecological relationships involving the tree may be uncovered. By visiting the tree at regular intervals, changes may be noticed and recorded.

Materials: A tree in a convenient location to observe throughout the year
A magnifier (optional)

Procedure: 1. Select a tree for study.
2. For the first visit, use the sample record form that follows for guiding observations.

FIRST VISIT TO TREE

A.   Identify a tree

    1. What kind of tree is it? _____

    2. How do you know? _____

    3. Where is the tree located? _____

B.   Description of tree's appearance

    1. Height (approximate) _____

    2. Bark _____

    3. Leaves _____

    4. Buds _____

    5. Fruit _____

    6. Branching _____

    7. Other _____

C.   Environment in which the tree is found _____

_____

_____

D. Economic uses of tree _____

_____

E. Relationships with other forms of life: insects, birds, plants, etc. _____

_____

3. Use this next record form for subsequent visits to the tree. These should be scheduled at regular intervals—weekly, biweekly, or monthly.

LATER VISITS TO TREE

A. General appearance of tree _____

_____

B. Change from last visit _____

_____

C. Changes on buds, leaves, ripening of fruit, flowers, bark _____

_____

_____

D. Any effects of rain, snow, or ice? _____

_____

E. Any evidences of visits by insects, birds, mammals?_____

_____

_____

Going Further: Drawings or photographic essays may be compiled of the tree to show the change of its appearance throughout the year. Some children may want to focus on one particular characteristic of the tree (bark, color, or insects, birds, or other animals living in the tree) and keep a diary of changes.

A careful study of a house plant may be undertaken in a similar fashion. Measurements and development of new flowers, leaves, or other growth can be recorded.

In determining the height of a tree, *triangulation*, a method used by surveyors to measure the distance to an inaccessible place, may be used (see pp. 196-198 for details on this fairly sophisticated technique).

STUDYING FLOWERS

Investigation: What are the main parts of a flower?   (I)

Background Information: Flowering plants are quite abundant and can be found almost everywhere on earth. They bloom in deserts, on mountaintops, in the water, and within jungles. Flowers contain the reproductive organs of the plant that bears them. There are two main types of reproductive structures in a flower—*stamens*, which produce pollen, and the *pistil*, which produces eggs or *ovules*. The color and the fragrance of a flower attract insects to it, and the insects (or wind in the case of some flowers) carry out the task of *pollination*. This essential process results in the fertilization of eggs deep within the body of the flower. Once the eggs are fertilized, they develop into seeds which then become enveloped in a *fruit*.

For classroom study, flowers with large parts, for example, tulips, lilies, or poppies, are most suitable and can easily be obtained at a local florist. If wild flowers are preferred, discretion should be exercised and rare flowers should not be taken. Among the list of common flowers which may be picked are aster, black-eyed Susan, bouncing Bet, buttercup, cinquefoil, clover, daisy, dandelion, milkweed, mullein, Queen Anne's lace, and yarrow, to name a few.

Materials: A few flowers with large parts, preferably tulips, lilies, or poppies
Magnifiers

Procedure: 1. Study the gross features of the flowers first, that is, stem, sepals (the green collar of leaflike structures which surround the flower), and petals (see Figure 1-5). Ask the children if they can guess the functions of these parts.

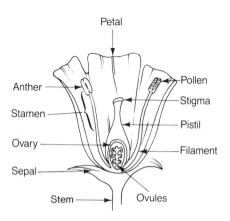

FIGURE 1-5.   A typical flower. The major structures contained in most flowers are shown here.

2. Gently pull the petals off the flower. Look at the pistil and stamens. How many stamens are there? Explain the function of each of these parts. The stamens are actually composed of two main parts—the *anther*, the top of the stamen which forms the *pollen* grains, and the *filament*, or the stalk upon which the anther rests. Touch the anthers. Does any pollen stick to your finger? The pistil is composed of the *ovary* (the swollen lower portion) and the *stigma* (the top of the pistil upon which pollen falls). Touch the stigma. It is usually sticky so that pollen will adhere to it.

3. If the pistil is large enough, it can be sliced in half (from top to bottom) with a razor, pin, or even a fingernail to examine the inside of the ovary and the eggs contained within it. Magnifiers may be needed to see the eggs (ovules).

Going Further: Models of flowers can be constructed using such materials as clay, toothpicks, tissue paper, and plastic. All parts of the flower may be represented.

Field trips to observe and smell flowers can be used to help children realize the importance of flowers in nature and to develop a respect for their existence. Reports on why wild flowers should be protected may be undertaken as a follow-up to such a field trip.

What is your state flower? Research may be done on why this particular flower was chosen. Almanacs and encyclopedias often contain this information. Can the state flower be found in the vicinity?

SETTING UP A TERRARIUM

Investigation: How can terrariums be set up and used as tools for learning?   (P, I)

Background Information: The terrarium is an important tool which may be used to help children acquire concepts of how living things coexist in a land habitat. Terrariums can be used to simulate such environments as a desert, a bog, grassland, and woodland. If a terrarium is well planned and carefully set up, organisms will thrive in it for long periods of time. For example, terrariums were used to transport plants from continent to continent in the days of sailing vessels when months were often required for the long sea voyage.

Terrariums may vary in size from a little jar with a few ferns, some moss or sand and a barrel cactus, to huge containers in which plants and animals live in close association. Frogs, toads, salamanders, turtles, and insects are often kept in terrariums along with plant life.

Through the use of terrariums, children can develop concepts concerning the requirements of plants, the adaptations which animals and plants have for living in their environments, and the nature and variety of plant and animal life.

Materials: A large glass or clear plastic container. An old aquarium which is no longer used because of a leak makes a fine container.

Various plants and animals according to the type of terrarium planned

Pebbles or small rocks

Sand

Soil

Charcoal (may be obtained at pet shops)

Procedure: The exact procedure followed will depend upon the type of environment to be simulated.

1. Put a layer of small rocks or pebbles on the bottom of the container.
2. Put some charcoal on top of the pebbles. This will absorb any noxious gases that may form.
3. Cover with the base for the type of environment to be created:

   For a desert terrarium—cover with 3 to 5 centimeters (1 to 2 inches) of sand. Insert small cactuses and a dead tree twig or two. Do not provide too much water, and keep the container uncovered. This terrarium should be kept near a sunny window.

   For a woodland terrarium—cover with soil and make a "hill" or slope within the container. A "pond" may be created at one end of the container (see Figure 1-6). Plant small ferns and tree seedlings, and put in some pieces of decaying wood. Keep moist and cover with a glass or plastic top to reduce evaporation.

For a grassland terrarium—cover with soil and plant grass seed. Keep moist and covered.

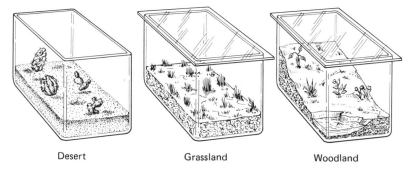

Desert   Grassland   Woodland

FIGURE 1-6.   Three types of terrariums illustrate the ways in which organisms live on land.

4. As children observe organisms growing in the terrarium, the following questions may be explored:

> Do plants grow about the same amount on each side of the terrarium? Is the growth greater on one side?
>
> Do the plants grow in the direction of the source of light?
>
> What happens when the amount of light changes? The amount of water?
>
> Do roots grow in the direction of a source of water? Do they tend to grow downward?
>
> What happens when roots encounter a pebble or some other obstacle?

Going Further: Terrariums can be used to study various interrelationships between organisms and between organisms and their environments. The following simple experiments, difficult to duplicate in the natural environment, can be carried out in terrariums:

1. What is the effect of temperature changes on plants and small animals such as insects? Keep a record of the temperature inside the terrarium for a few days. Then place a small heat source near the terrarium. What changes occur after a few days? Try not to permanently damage the plants. What happens when the heat source is removed?

2. What is the effect of changes in the amount of light on plants? First expose the terrarium to light around the clock. What happens after a few days? Then cover the terrarium so that it receives no light. What happens when the terrarium is returned to normal light conditions?

3. What is the effect of changes in humidity on plants and animals? Except for the desert terrarium, a glass plate should cover the top of the tank to prevent the loss of moisture. Do drops of water form on this glass plate? If so, where do the drops come from? Remove the glass plate for a few days. Note any changes.

4. What happens when plants grow so that the leaves touch the sides or the top of the terrarium?

ANIMALS IN THE NEIGHBORHOOD

Investigation: Which animals live in the local community?   (P, I)

Background Information: Youngsters may be surprised to learn how many animals they can find living in their community. Areas for study should be defined in advance, and children should be instructed to focus on and look for animals in a particular spot. Identifying the specific names of the animals found is not important during an initial visit to the local community.

Materials: Small shovels, trowels, or spoons for digging in soil
Plastic bags for collecting animals
Magnifiers
Meter stick
Pencils and pads

Procedure: 1. Find a nearby spot where there is a tree, a concrete sidewalk, and soil.
2. Assign groups of children to look for animals in specified areas—within a square meter of soil, on the sidewalk, in and around a tree, above their heads.
3. Distribute the appropriate materials required for exploration in each area.
4. Ask the youngsters to draw pictures of the animals they find. If they know the names of them, they should list them.
5. Collect samples of animals which may seem appropriate for further study—for example, worms, bugs, millipedes, centipedes, ants, and so on.
6. After returning, list the types of animals discovered and the environments in which they were found.

Going Further: Particularly motivated youngsters can be given animal guides or books and asked to identify the specific names of the speciments collected. Some children might want to research and write a report on one of the animals. Which animals found were the smallest? Which were the largest? Which animals were the most colorful? The least colorful? Which animals normally live in the soil? In trees? In water? Which animals have the most legs? The fewest legs? Which animals move the fastest? The slowest? What do these animals actually eat? Are they found in all seasons? How many may be kept and cared for in the classroom?

COLLECTING AND STUDYING INSECTS

Investigation: How can we collect insects, build cages for them, and study them? (I)

Background Information: Insects are of tremendous economic importance to humans. Their small size, enormous variety, and rapid rate of reproduction have made them a very successful animal group. There are nearly one million different types (species) of insects! You can only imagine how many billions of individual insects exist in the world. Many varieties of insects are useful to us in that they feed on plants and animals that are harmful; however, insects in themselves are often harmful in that they carry diseases. Some medicines are made from insect bodies, as are shellacs and dyes.

All insect bodies have three main parts: head, thorax, and abdomen. They have three pair of legs, one pair of antennae, and most often one or two pair of wings. Insects have inhabited almost every conceivable environment on the earth. They exist in jungles, deserts, frigid zones, and on mountaintops. Among the long list of insects are ants, bees, wasps, dragonflies, fleas, flies, bugs, grasshoppers, termites, beetles, moths, and butterflies, to name a few. The supply of insects for classroom study is readily available in all seasons but winter.

Materials: A net (may be easily made from the toe of a nylon stocking, a coat hanger, and a broom handle)

Glass jars

Soil

Sod or some fresh grass or clover

Procedure: 1. Collect some insects with a net. Grasshoppers are relatively large and seem best suited for classroom study.

2. Place some soil and sod (or fresh grass or clover) on the bottom of a jar and moisten (see Figure 1-7).

FIGURE 1-7. A jar with a mesh or stocking top is a suitable environment for keeping insects.

3. Place the insects inside the jar and cover the top with a piece of nylon stocking. Secure the stocking to the top of the jar with a rubber band or string.

4. From time to time put small pieces of sweet apple or melon in the jar to feed the insects.

5. The children may observe, draw, and possibly witness the various stages in the life cycles of insects. It will be interesting to compare the similarities and differences among various insects collected. Can the three main body parts be identified?

Going Further: If crickets are found and kept in the classroom, interesting studies may be conducted on the number of chirps heard in a minute. By placing the cricket cage in a sunny window or on a cool shelf, differences in the chirping rate will be noticed. To check this relationship between the number of chirps and the environmental temperature, count the number of chirps in one minute, divide by 4, add 40, and the result should be the approximate temperature near the cricket. Check with a thermometer.

Most insects pass through various stages of a life cycle. Some children may wish to research the life cycle of a particular insect and report to the others on it. Insects can be mounted and displayed for exhibits. Butterfly collections can provide many hours of interest for motivated youngsters.

ANTS

Investigation: How can ants be collected, kept, and studied? (P, I)

Background Information: Studying the ways of ants is a fascinating activity for children to engage in. Ants are found nearly everywhere—in backyards, school grounds, gardens, or sidewalks. Ants live within a colony, or a maze of interconnected tunnels. Displaying a high degree of social organization, ants depend upon one another and exchange food materials among each other. The queen ant, who is larger than the other ants, can fly during mating. The queen's purpose is to maintain the colony by producing young. The workers, the most populous subdivision of ant society, carry in the food for the colony. They predigest it for the larvae (newly hatched individuals). The larvae, in turn, produce fluids which the workers like. Some ant species have a soldier caste that protects the colony from other insect invaders. Ants have highly developed senses which help them keep track of where their nest is when they go off in search of food. Keeping an ant colony in the classroom can provide unique opportunities for youngsters to observe the life of ants.

Materials:
A large, wide-mouthed glass jar

A large can (from canned food) which can fit inside of the glass jar, but not leave too much space between it and the jar

A garden trowel or small shovel

Soil

A supply of ants

Food for the ants—bread crumbs, jelly, and sugar

A loosely woven cloth to cover the jar

Black construction paper or aluminum foil large enough to surround the glass jar

Procedure:
1. Place the sealed can inside of the glass jar so that there is no more than 3 or 4 centimeters (1½ inches) of space between the can and the sides of the jar. This requires the ants to stay close to the glass as they build their tunnels, and thus makes them more visible to the children (see Figure 1-8).
2. Sprinkle the soil around the can until the jar is filled.
3. Try to find an ant hill. Dig down with a trowel or shovel to remove the entire colony of ants. Try to get the queen (a much larger ant), as she is required for the maintenance of the colony. (Ants for classroom use are also available through scientific supply houses or as complete kits with an appropriate cage in toy stores.)
4. Place the ants in the jar. Also include some bread crumbs, sugar, and a little bit of jelly for food.

FIGURE 1-8. An ant colony can be maintained in a jar.

5. Place a wet sponge on top of the soil (for moisture), and cover the jar with a loosely woven cloth secured with a rubber band.

6. When no one is observing the ants, cover the outside of the jar with black paper or aluminum foil so that the ants can avoid light as they build their tunnels.

7. Once the colony is established, regular observations of its social activities can be made. Children can see how ants carry particles, how they eat, and how they rest.

Going Further: Experiment with different types of food for the ants. How do they respond to lettuce, a tiny piece of chopped meat, bread crumbs? What conclusions can be drawn about the diet of ants?

Many observations can be made on a sidewalk or near an anthill. How many different kinds of ants are seen? How many legs do they have? Do they have wings? Can their eyes be seen (use a magnifier for this). How does the size of the ant's load compare with the size of its body? What happens when two ants meet? If an obstacle is put in the path of an ant carrying a particle, what happens?

COCKROACHES

Investigation: How do cockroaches live and change?   (P, I)

Background Information: Despite their widespread, and undeserved, unpopularity, cockroaches are very interesting insects. Fossil evidence reveals that cockroaches have remained very much the same for hundreds of millions of years. They belong to a large group of insects called *straight-winged insects* and are related to locusts, grasshoppers, and crickets. The female cockroach lays a small, bean-shaped pouch, which if placed in a grocery bag, can unknowingly be carried home—thus, it is quite possible to find a cockroach in even the most immaculate home! Roaches like dark, moist areas and tend to be most active at night. While they do not transmit any specific disease, they probably can carry germs onto any food they touch. They eat many of the same kinds of foods that humans eat and will also eat glue from postage stamps and book bindings.

Like crickets and grasshoppers, cockroaches undergo *incomplete metamorphosis* (that is, after hatching, the insect will always have the appearance of the adult). The eggs hatch into tiny cockroaches which shed their outer coverings (molt) several times before they reach maturity. Adult cockroaches have wings, and some can fly short distances.

Materials: A wide-mouthed glass jar

A piece of cardboard or stiff paper

Potato peel or any other moist food

Procedure: 1. First try to capture a cockroach. They can often be found in dark, wet places. It may not be possible to find roaches on the day you want to accomplish this activity, so it is wise to keep the needed materials on hand for just the right moment. When cockroaches are seen, place a wide-mouthed jar over them and slip a piece of cardboard or stiff paper under the mouth of the jar. Quickly turn the jar over (see Figure 1-9). Moist paper should then be placed in the jar so

FIGURE 1-9. To capture a cockroach, place a wide-mouthed jar over it. Slip a card underneath it and then quickly turn the bottle over.

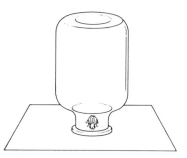

that the cockroaches will be able to retreat into dark areas within the folds of the paper. Punch holes in the jar's lid and screw it back on the jar.

2. Have the children observe the cockroaches, the paper, and the jar over time. Potato peels or any other moist food may be fed to them. What changes take place?

3. Cockroach droppings look like small black balls. Young cockroaches hatch out of small egg pouches and are extremely small at birth. Can any be seen?

4. How do the cockroaches seem to use their antennae (the stringlike feelers attached to their heads)? Can wings be seen on the cockroaches?

Going Further: Particularly interested children may research the history and habits of cockroaches. Their findings may be presented to the class. There are two main types of cockroaches in North America: the German and the American cockroaches. What are the differences between these two varieties?

If placed in a sealed, clear-plastic container, cockroaches can be studied and observed on a screen while the container rests on the glass of an overhead projector. Experiments can be conducted to determine whether the cockroach prefers light or dark, moisture or dryness, and different types of food.

BUILDING A WORMERY

Investigation: How can a wormery be constructed and what observations of worms can be made?   (P, I)

Background Information: Earthworms are *segmented* worms. Their bodies are arranged as a series of similar segments. Earthworms live in moist soil, burrowing through it as they ingest the decayed plant and animal matter in it. A mouth opening can be observed at one end of the earthworm, and each segment has bristles attached to it. These bristles help the earthworm to move about on the surface of the soil or through it in its underground burrows. Earthworms make soil more fertile by loosening it as they move through it.

Earthworms are relatively easy to find. After a heavy rain, and before the water has had time to drain away, earthworms can be seen coming out of their burrows. Digging in moist soil or turning over a rotting log or rocks will usually reveal many of these little creatures.

Materials: A trowel or shovel
A large jar or plastic food container
Soil mixed with decaying leaves
Cornmeal (to feed the earthworms)

Procedure: 1. Obtain a container for the earthworms. A large glass jar or plastic food container will do.
2. Fill the container three-fourths full of soil that has been mixed with decaying leaf material.
3. Collect a dozen or so worms by digging in moist soil. Earthworms may also be brought to the surface by driving a stake into the ground and then rubbing a board over the top of it. These vibrations bring the worms out of their burrows.
4. Put the earthworms in the container. Keep the soil moist by covering the container with a wet cloth.
5. For food, place a handful of cornmeal in one corner of the container. Do not feed more than twice a week. The cornmeal may occasionally be combined with some coffee grinds.
6. Many observations may be made of the earthworms. Do the two ends of the worms seem exactly alike? Do the worms prefer light or darkness? Do they prefer moist or dry environments? Do they prefer warm or cold spots? Children may devise investigations to answer each of these questions.

Going Further: An earthworm can be placed in a clear tray or baking dish resting on an overhead projector. Observations of the earthworm's movements can then be projected on a screen for a group of children to observe. If dark construction paper is placed underneath one half of the dish (thereby blocking the light from that portion of the dish), where does the earthworm go? Does it tend to avoid the light? Other such experiments can be conducted as the children view the results on a screen.

Obtain two plants of the same variety and similar size. Keep them in the same size containers and in the same soil mixture. Put some earthworms into the soil of one pot and none in the other. Maintain the same lighting and watering conditions for both plants. Observe the plants for a few weeks. Is there any noticeable difference in the rate of growth and condition of the plants? If so, what might account for it?

SETTING UP AN AQUARIUM

Investigation: How can an aquarium be set up and used as a tool for teaching and learning? (P, I)

Background Information: Aquariums generally are simple to set up and require careful attention but relatively simple maintenance procedures. Classroom aquariums offer many unique opportunities for teaching and learning. Children can acquire many important concepts concerning the needs of aquatic plants and animals through their experiences in establishing and maintaining a balanced aquarium. The interdependence of plants and animals may be demonstrated clearly in an aquarium. Plants produce oxygen which is used by the animals, and animals in turn give off carbon dioxide which plants use during *photosynthesis*, the process by which they manufacture food. Plants may also be a source of food for the animals in an aquarium.

Most often, it is best to purchase plants, fish, and snails in a pet shop, but some teachers may want to stock aquariums with life from a local pond or brook. Aquarium tanks vary in size, cost, and complexity of available options and gadgets. It is advisable to start with a simple setup in a classroom. A gallon jar is probably the simplest container to start with; a 5-gallon or 10-gallon tank is even better.

Materials: A container of at least 1-gallon capacity. A 5-gallon or 10-gallon tank is preferable.

Clean, coarse sand sufficient to cover a 5 to 10 centimeter (2 to 4 inch) layer on the bottom of the container.

Aquatic plants from a small pond, stream, brook, or an aquarium supply store. Vallisneria, Sagittaria, and Elodea are common aquarium plants.

Fish, snails, and tadpoles. Goldfish, mollies, and guppies are usually the most hardy fish available in pet shops or department stores.

Procedure:
1. Thoroughly wash the container to be used. The glass should be clean to facilitate observation.
2. The coarse sand should be washed and stirred several times until the waste water is completely clear. Place a cover of 5 to 10 centimeters (2 to 4 inches) of sand in the container (see Figure 1-10).
3. Place a piece of paper over the sand and pour in clean water to a height of 10 to 12 centimeters (3 to 4 inches). Remove the paper. Wash the plants and set them into the sand. Make sure the roots are entirely covered with sand.
4. Fill the aquarium with water to within 3 to 5 centimeters (1 to 2 inches) of the top. Do not use water drawn from copper pipes, as such water may be harmful to the fish. If possible, use spring water

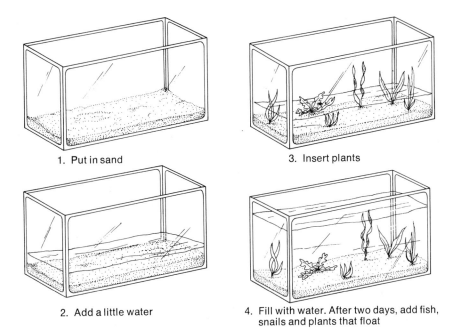

1. Put in sand

3. Insert plants

2. Add a little water

4. Fill with water. After two days, add fish, snails and plants that float

FIGURE 1-10. Setting up an aquarium helps show children some of the ways that organisms live in water.

which is available in most supermarkets. Allow the water to stand for two days to allow any chlorine dissolved in it to escape.

5. Place the water animals (fish, snails, tadpoles) into the tank. If the fish seem to swim consistently near the surface of the water, there may not be enough oxygen in the tank. Additional plants should be added, or a small pump with an aerator may be needed. Keep the aquarium in a light spot, but away from direct sunlight.

6. Do not feed the fish more than 2 or 3 times per week, and never feed more than is eaten in 10 minutes.

7. The aquarium should be covered with a glass or plastic pane. This reduces evaporation and prevents dust and dirt from contaminating the aquarium.

8. The following observations of fish may be made:

   Observe the actions of the fins. How does each aid in navigation?

   Find the operculum (gill cover). How many times does it open and close in a minute? What happens if the fish is placed in warmer or cooler water?

   What differences can be observed among the fish in the aquarium?

   Can any "pecking order" among the fish be observed?

Can fish learn? Tap the sides of the aquarium before each feeding. Do the fish learn to come to the top of the water in search of food whenever the aquarium is tapped?

9. The following observations of snails can be made:

Look at the thin stalks at the front of the snails. What happens if they are tapped lightly with a pencil?

Look at the snails' movements across the aquarium glass. What steps are involved in the movement?

Place a piece of glass or plastic over a sheet of graph paper. Have a snail move over the length of the glass. What is the snail's rate of movement in kilometers (miles) per hour? What is meant by the expression "moving at a snail's pace"?

10. Where do air bubbles appear in the aquarium? How could these bubbles have been formed? What could be the reason for this?

Going Further: After the children have had some experience in maintaining a simple aquarium, it may be desirable to have them set up and maintain a more sophisticated aquarium. A pump will help insure that sufficient air will be dissolved in the water. There are two main types of aquarium pumps: vibrator pumps and the considerably more expensive piston-type pumps. Filtering systems and aerators are usually attached to the pump. A thermostatically controlled heater makes it possible to introduce tropical fish and other organisms that are sensitive to heat changes. Saltwater aquariums represent a particular challenge for an especially motivated group of youngsters.

MAKING A BIRD FEEDER

Investigation: How can a simple bird-feeding station be built and maintained? (P, I)

Background Information: Birds are the only animals that have feathers. They are warmblooded and have four-chambered hearts not unlike our own. Their bodies are intricately adapted for flight; for example, their bones are hollow, and they have air sacs attached to their lungs. This results in a comparatively light body weight which is necessary for flight.

Bird watching is an interesting activity and a lifelong avocation for many people. This interest can be introduced and developed in the elementary school classroom. The bright colors and songs of birds are prime examples of beauty in nature. Children should be encouraged to observe birds in parks, around schools, and near their homes. After a few local bird species are seen repeatedly, children will begin to recognize them and name them by sight or by their song.

Materials: A piece of board (approximately 30 centimeters by 30 centimeters, or one foot square)

Wooden molding, enough to go all around the edges of the board

Small nails and a hammer

A wooden post if the feeder cannot be secured to a window ledge or a tree trunk

Binoculars (optional)

Bird guides

A small dish for water, and seeds, crumbs, grains, and cereal for food

Procedure: 1. Nail the wooden molding to the edges of the wooden board so as to form a lip around the board to prevent food from blowing away (see Figure 1-11).

2. Secure the board to the outside ledge of a classroom window, on a post, or to tree branches within sight of the classroom window.

3. Place a water dish on the feeding platform, as well as grains, dry cereals, all types of seeds, cracker crumbs, and the like. All of these foods may be rolled into a ball using peanut butter as the "glue."

4. As birds visit the station, the children should be careful not to make sudden movements or loud sounds, as these will frighten the birds away.

5. As birds appear, the children should look them up in bird guides and manuals and identify them. A list should be kept of the dates and the visitors to the station.

6. The following observations of birds may be made:

What type of bill does the bird have? Does the bill seem to be specially adapted to a particular feeding habit?

What do the bird's feet look like? Look for scales on the feet. Do the bird's feet seem to be adapted for a specific purpose?

Observe the bird's tail. What does it seem to be used for? Describe the shape of the wings.

Does the bird fly in a straight line, like a roller coaster, or does it soar?

When the bird walks, does it hop, walk, or jump?

Does the bird call or sing? Is the song recognizable?

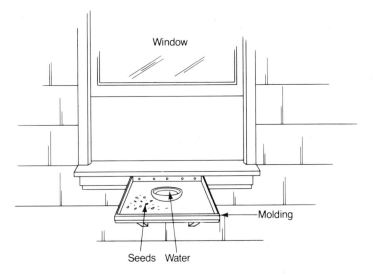

FIGURE 1-11. A simple bird-feeding station may be constructed using a wooden board and some molding.

Going Further: Take a walk through the community. Can any birds be seen? Can they be identified? Are any of their calls or songs familiar? What is your state bird? Can it be seen in or near the school grounds? Some children might want to prepare a report on their state bird.

Serious bird watchers maintain a "life list." This is a cumulative record of the names, dates seen, and locations of birds observed. Children might want to initiate their life lists as part of a class or group project.

The migratory habits and patterns of birds are fascinating topics for study. Do any of the major migration flyways pass through the community in which the children live? How might they find out? Are the dates of migration predictable?

KEEPING MAMMALS

Investigation: What observations can be made of mammals? (P, I)

Background Information: Mammals are perhaps the most popular classroom animals. Gerbils, hamsters, mice, rats, and guinea pigs are relatively easy to care for and observe. Keeping animals also provides opportunities to establish respectful attitudes on the part of the youngsters and to assume new responsibilities. Large glass or plastic aquariums (even old ones that have leaks) make fine cages for these animals. In most cases the groundcover should be cedar shavings mixed with some chlorophyll chips to absorb odors. Gerbils, hamsters, mice, rats, and guinea pigs may all be fed a mixture of dry food pellets and sunflower seeds (particularly for gerbils), with occasional strips of lettuce and carrot.

Before a pet is obtained, the children in the group should thoroughly research the physical needs of the animal, including dietary, environmental, and handling requirements. When the pet actually is brought to the classroom, procedures for its care (and the supervision of those caring for the animal) should be clearly defined. Care during weekends and vacation periods should be arranged in advance. This can provide youngsters who do not have pets at home with an occasional rare treat. But please secure parental permission for this visit! Booklets on the care and feeding of any classroom pet should be close at hand and provided to vacation or weekend caregivers.

Materials: The animal to be kept or observed (Make certain that any pet purchased, acquired, or "adopted" is in good health and free from disease.)

A suitable cage or environment for the animal

Appropriate ground cover and food

A water bottle

Manuals and guides on the care and feeding of the pet

Procedure:
1. Discuss the needs and requirements of the pet to be acquired. Manuals and guides should be close at hand and referred to whenever necessary. Procedures for the daily chores to care for the animal can be posted on a chart near the cage or environment. Establish clear directions for the animal's care. (A teacher of 25 children would not want the animal fed 25 times each day!)
2. Have the animal checked by a knowledgeable individual to insure that it is in good health. Obtain an adequate food supply and appropriate ground cover.
3. The following observations may be made of the pet:

    What foods does the pet seem to enjoy most? Does it eat at a particular time of day? Does the animal clean its paws after eating?

The animal may be weighed at regular intervals to check on its growth. (Weigh a box first, then place the animal in the box on a scale. Which mathematical operation is needed to determine the animal's weight?) A graph may be plotted of the animal's weight changes.

A log may be kept of the animal's daily activities. When does it sleep? When does it appear to be most active? Does it interact with a mate or other animals? Children may wish to focus on a particular aspect of the animal's activity, for example, eating, play, mating, reactions to a flashlight or other harmless stimuli.

A maze may be built for the animal to run. Does the animal learn the way out of the maze after a few trials? If a T-maze is used (that is, one in which there is an initial pathway and then a branching to the right and left), does the animal respond to a reward (food) at one end of the maze?

Note: Animals should never be deprived of food or subjected to any harmful stimuli.

Going Further: Reports of various aspects of the animal's life may be researched and presented to the class by an individual or a committee. Some interesting topics might include the history and origin of the animal, its closest relatives in the animal kingdom, its dietary preferences, results of learning experiments, its reactions to stimuli, mating and reproductive habits, varieties of the animal which are different from the classroom pet, and its relationships with other animals in nature.

2

Air, Water, and Weather

INTRODUCTION

"Everybody talks about it, but no one does a thing about it." That is an often-heard expression concerning weather. Atmospheric conditions and climate do have important effects upon our daily lives. We wear certain kinds of clothing because of the weather. We heat, air condition, and insulate our homes because of temperature extremes. In schools, graduation exercises are suddenly moved due to rain, class trips are cancelled, and recess activities are held indoors—all because of adverse weather. But weather need not always be considered a hindrance. Weather conditions often evoke strong human emotions, feelings of well-being and rebirth, as many poets have attributed to the spring. Few experiences can be as exhilarating as a crisp autumn morning in northern North America, a gentle breeze sliding past the face, or the first balmy day after a severe winter.

Weather is caused by the interaction of the sun, air, water, and land. The sun warms the earth, but it does not warm all places equally. Some regions of the earth (namely those in the tropics) receive the sun's most direct rays, that is, the rays that strike the earth's surface at nearly right angles. The land is heated, and so too is the air immediately above. This sets up a climatic cycle in which the warm air rises, then becomes cooler, resulting in some precipitation of the moisture held within it, which is then deflected to other areas of the earth. Different earth materials (for example, rock, soil, sand, water) are affected differently by the sun's warmth. Day and night also affect how these materials are heated. Some absorb and hold heat, while others are subject to considerable daily temperature fluctuation. This uneven heating of the earth is primarily responsible for our weather.

In order for youngsters to understand the major factors that govern our weather, they should first investigate the properties of water and air and the interaction of these two elements. The activities that follow, as well as some in the next chapter, represent a first step toward appreciating the important relationships among air, water, land, and the sun which account for our daily weather.

WATER IN ITS THREE STATES

Investigation: How can the three states of matter be demonstrated with water? (I)

Background Information: Water changes from solid to liquid to gas at temperatures that can be produced fairly easily. Most substances exist naturally on the earth in one, or at most two, of these three states of matter. Heating, cooling, or compressing can cause substances to pass from one state to another. However, water is unique in its natural abundance in all three states of matter on earth. As ice, it is a *solid*. Much water is in the form of ice. Greenland and Antarctica are mostly covered with ice. During the glacial periods, much more of the earth was covered with ice. Water as a *liquid* covers most of our planet in oceans, lakes, rivers, and so on. Water vapor is present in the air, and in this form it is a *gas*.

As mentioned in the introduction to this chapter, our weather is caused by the interaction of the sun, air, water, and land. So, as children begin to engage in activities involving weather, it is important that they first have some concrete experiences with water and its properties.

Materials: An empty food can

Ice cubes

A thermometer with a range that starts below 0° C (32° F) and reaches above 100° C (212° F)

An electric hot plate

Procedure: 1. Place ice cubes in an empty food can and mix with a little bit of water. Stir the mixture of water and ice. Place a thermometer into the mixture before all of the ice has melted and record the temperature. This is the melting point, and the thermometer should read 0° C or 32° F. (Some water beads may be present on the outside of the can. Ask the children where they think these droplets have come from. Actually, they are small drops of water vapor from the air which have condensed into liquid water from the coolness of the outside of the can.)

2. Place the can on an electric hot plate and heat it until the water boils. Insert the thermometer and record the temperature again. This is the boiling point, and the thermometer should read 100° C or 212° F. In boiling, the water changes from its liquid to its vapor or gaseous state. Water vapor is actually invisible. The small clouds that we see when water boils are composed of small water droplets that have condensed from the vapor.

3. Change of state in the opposite direction can also be demonstrated. Water vapor will condense into liquid on a cool surface, such as a pan

filled with water. The liquid water that is collected can, of course, be frozen again in the freezing compartment of a refrigerator.

4. As a result of this activity, ask the children if they can see any benefits of using the Celsius system as opposed to the Fahrenheit system of temperature measurement.

Going Further: There are other properties of water which children may wish to investigate. Color is a property of matter. What is the color of water? Odor is another property. What is the odor of water? Is water tasteless?

Water is a solvent. Many kinds of materials are easily dissolved in it. Have the youngsters try to dissolve various substances in water, for example, sugar, salt, coffee lightener. Ask the children to list those substances which can be dissolved in water and those which cannot. Water's ability to hold dissolved materials can be compared with that of another liquid. Place equal amounts of water and kerosene into separate glass jars. Put a spoonful of sugar into each of the liquids and stir. In which liquid does the sugar dissolve?

Water is a heavy liquid. Pour water and kerosene to equal heights in two tall jars, such as olive jars. Float a pencil in each of the liquids. In which of the liquids does the pencil float highest? What does this mean?

Ice and water have slightly different densities. Place an ice cube in a jar of water. Push the ice cube down to the bottom of the water and release it. What happens to the ice cube? Water expands when it freezes and becomes less dense (see pp. 43-44). In the winter, lakes and streams are covered with ice and seldom freeze to the bottom.

WHEN WATER FREEZES

Investigation: What happens when water freezes? (P, I)

Background Information: An important characteristic of water is that it expands as it freezes. This phenomenon has important implications for life as we know it. Since water expands as it freezes, it becomes less dense. The same amount of material occupies more space; therefore, it becomes lighter for each unit of volume. (A cubic meter of water weighs more than a cubic meter of ice.) Being less dense, ice rises to the surface of a pond or lake at subfreezing temperatures. The formation of ice on the surface of a pond or lake protects the waters beneath it from exposure to cold air, so that living things do not freeze. If bodies of water froze from the bottom up, many would never thaw during the summer, and we would have much colder climates. Under such circumstances, living things would have to be very different in order to survive.

Ice is an important geologic agent, especially in temperate and arctic regions and on mountaintops. Water seeps into rock crevices and pores during mild periods. When this water freezes, it expands, and is capable of cracking rocks. This process is an important factor in the disintegration of rocks and the formation of soil.

Many children have had experience with the pipes in their homes bursting during a cold spell in the winter. This is again due to the fact that water expands as it freezes. There is not enough room in the pipes, which were previously filled with liquid water, to hold the frozen water. In this investigation children witness, in a concrete way, the expansion of water as it freezes.

Materials: A small jar with a screw-on lid

A paper or plastic bag large enough to hold the jar

Procedure: 1. Fill a jar of water to the very top and screw the lid on very tightly.
2. Place the jar of water inside the paper or plastic bag.
3. Place the jar and the bag in the freezer compartment of a refrigerator.
4. Remove the jar the next day. What happened? How do the results demonstrate that water expands as it freezes? How can this experience be related to the bursting of water pipes as the water within them freezes in some climates?

Going Further: This phenomenon can be explored using other methods and materials. Fill two juice cans or similar containers with equal amounts of water. Place one in a freezer (or outdoors if it is cold enough), and leave the other at room temperature. After the water in one of the containers has frozen, compare the height of the contents. Do the containers still seem to be filled to the same level? What other factor, aside from the expansion of water, may have influenced the results?

Fill two ice cube trays to the same level with water. Place one in the freezer compartment of a refrigerator and the other in the refrigerator itself. After the water in the tray in the freezer compartment is completely frozen, compare the height of the tops of the ice cubes with the height of the water in the other tray. What accounts for the difference?

An interested youngster may wish to devise an experiment to demonstrate how water, when frozen, can actually crack a rock. In order for this experiment to work, careful selection of the rock and conditions of freezing have to be considered.

BRICKS SOAK UP WATER

Investigation: How much water will a brick soak up?   (P, I)

Background Information: Rocks, bricks, and other materials in the environment are porous and can absorb water. An interesting set of calculations which children might like to work with determines the amount of water which bricks or rocks can soak up. This procedure requires that children apply skills in volume measurement and subtraction.

   The fact that rocks are porous and can absorb water is an important part of the soil-making process. If a rock absorbs water during a rainfall and is suddenly subjected to freezing temperatures, the water in the rock will expand as it freezes, making the pores larger, weakening the rock, and eventually cracking it (see pp. 43-44). Repetition of this process over long periods of time results in the formation of soil.

Materials: A tank or basin

A large measuring cup or container with volume markings on it

An assortment of rocks and bricks

Procedure: 1. Put a brick in an empty tank or basin.

2. Pour a measured quantity of water into the tank or basin sufficient to completely cover the brick.

3. Leave the brick in the water for about half an hour. What do the children see happening? What might account for these observations?

4. Remove the brick and blot it dry with a cloth or paper towel.

5. Pour the water left in the basin back into the measuring cup or marked container.

6. Ask the children to calculate how much water the brick soaked up. What conclusions can be drawn about the porosity of the brick?

7. Repeat the procedure using various kinds of rocks. Which types of rocks seem to soak up the greatest amount of water? (Before definitive comparisons can be made, the children will have to consider the sizes of the rocks they are comparing.) Do these results have any implications for the process of rock breakdown? Which rocks would be more resistant to weathering?

Going Further: There are other calculations which children may wish to make using water. For example, how much does a liter of water weigh? How much does a gallon of water weigh? If a ten-gallon fish tank were filled to the top with water, how much would it weigh? (Do not forget the weight of the fish tank itself!) If the room that the children are in was filled with water, how much would it weigh?

A WATER-DROP MAGNIFIER

Investigation: How can a magnifier be made from a drop of water?   (P, I)

Background Information: An important property of water is its ability to bend light which passes through it. If a stick or pencil is placed inside of a glass of water, it will appear broken because the image we see is altered by the bending of light rays as they pass from one medium (air) to another (water) (see pp. 79-80.

A drop of water can act as a magnifying lens and make objects appear larger than they actually are. In this easy-to-do investigation, children can witness this interesting property of water by making a water-drop magnifier.

Materials: Glass microscope slides

A grease pencil or china marker

An eyedropper

Procedure: 1. Using a grease pencil or china marker, draw a circle about 5 centimeters (1/4 inch) wide on a clear glass microscope slide.

2. Using an eyedropper, carefully place a drop of water on the slide so that it rests inside of the circle.

3. Slip a small piece of newspaper under the slide. Do the letters appear larger when viewed through the glass or through the drop of water? Why does this work? How does the circle drawn aid in keeping the drop of water intact? (See Figure 2-1.)

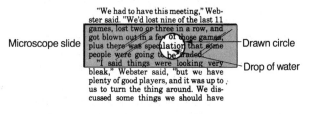

FIGURE 2-1.   A water-drop magnifier can be made from a microscope slide and a drop of water.

Going Further: Using various small containers of water, children can find other ways of making a water magnifier. If a test tube with a stopper is available, fill it with water and plug it. Can the test tube be used as a magnifier? Can a small flask or clear medicine bottle be similarly used as a magnifier?

The magnifying power of a water magnifier or a glass magnifying lens can easily be determined. Hold the magnifier in focus over a piece of lined paper. (The lines should be fairly close together, so graph paper may work better than lined paper.) Compare the number of lines seen outside of the lens with a single space as seen through the lens. This ratio is the power of the magnifier.

A WATER CYCLE

Investigation: How can a water cycle be demonstrated? (P, I)

Background Information: Water on the earth is involved in a constant circulation called the *water cycle*. Water evaporates from soil, the leaves of plants, the lungs and skin of animals, and from puddles, ponds, lakes, and other bodies of water. In its vapor state, water is carried through the air and eventually condenses into small droplets to form clouds, and from the clouds, the water falls back to the ground in the form of rain or snow. When the rainwater enters the ground, it is again available for use by plants and animals, and the cycle continues (see Figure 2-2).

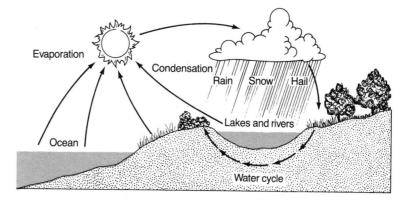

FIGURE 2-2.   A diagram of the water cycle.

Children can replicate a water cycle on a very small scale using simple materials. From this demonstration, the youngsters can gain a better understanding of how the water cycle operates on a larger scale and how it is an important factor in our daily weather.

Materials: A plastic bag
A twist-tie for the bag

Procedure: 1. Pour a small quantity of water into a plastic bag. (A cup of water is more than enough.)
2. Seal the bag with a twist-tie.
3. Place the bag on a windowsill in direct sunlight. Observe the beads of water forming on the top and sides of the bag. How did they get there? (See Figure 2-3.)

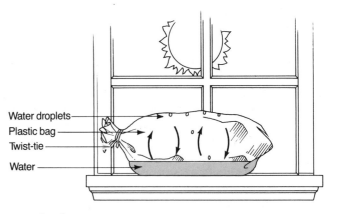

Water droplets
Plastic bag
Twist-tie
Water

FIGURE 2-3.   A water cycle on a small scale.

4. Place the bag in a cool place away from the sunlight. What happens to the beads of water? Why does this occur? How does this demonstration explain some aspects of our daily weather?

Going Further: There are other cycles which occur in nature which the children may wish to research or investigate. Oxygen and carbon dioxide are involved in the *oxygen-carbon dioxide cycle.* Green plants use carbon dioxide to manufacture food, and in this process (called photosynthesis), oxygen is released. Both plants and animals use oxygen to oxidize food to release energy, and in this process, carbon dioxide is released again.

Nitrogen is an essential element involved in another cycle, *the nitrogen cycle.* When plants and animals die and decay, small organisms promote this decay and convert the nitrogen in the dead bodies of the organisms into a form that can be used again by growing plants, and so the cycle continues. A group of tiny organisms called nitrogen-fixing bacteria can take nitrogen directly from the air and form nitrates which can be used directly by plants. Children may enjoy diagraming these cycles, producing setups to demonstrate them, or identifying other natural cycles.

WHEN WARM MEETS COOL

Investigation: How can the interaction of two air masses be demonstrated?   (I)

Background Information: When objects are heated they expand (see Chapter 3). As air is heated, it becomes less dense, and therefore rises. In weather reports, reference is often made to hot and cold air masses. In order to better understand how such air masses operate in nature, it is possible to produce simulations which in essence replicate natural phenomena on a small scale.

In this activity, children can simulate the meeting of warm and cold air masses. The weather that might be experienced by people on the ground can then be predicted.

Materials: Two milk or ketchup bottles
Food coloring
An index card

Procedure: 1. Fill one bottle half way with hot water. Add a few drops of food coloring. Then completely fill the bottle with hot water to the top.
2. Fill the other bottle with cold water.
3. Place the index card on top of the bottle of hot water. Keeping the card pressed firmly to the mouth of the bottle, quickly invert it over the mouth of the bottle of cold water. (It is best to do this over a shallow pan in the event that some water spills.)
4. Pressing the two bottles firmly together, carefully turn the two bottles on their sides so they are lying, mouth to mouth, on a table top (see Figure 2-4).

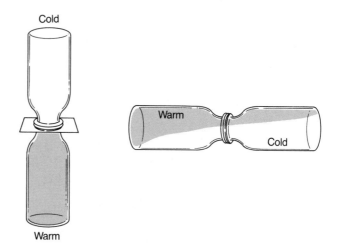

FIGURE 2-4.   What happens when warm water meets cold water?

5. Remove the card and be sure that the mouths of the bottles are in contact so that leaking does not occur.

6. What happens to the water in the two bottles? Where does the warm (colored) water eventually lie with respect to the cold water?

7. If the hot water represents a warm air mass and the cold water represents a cold air mass, what happens to the weather on the ground when a warm air mass meets a cold one?

Going Further: In some large bodies of water, currents of warm water meet currents of cool water. How can the above investigation demonstrate what happens when these currents meet? Often in nature a body of salty water meets a source of fresh water. If salt water is placed in one bottle and fresh water is placed in another, and food coloring is placed in either one, the effects of salt and fresh water meeting can also be simulated. Ask the children if they can identify areas in the world where this situation occurs.

Occasionally, warm air gets trapped under a mass of cooler air, leaving pollutants (such as industrial wastes, automobile exhaust fumes, and smoke from incinerators) in close contact with the population on the ground. This situation, called a *thermal inversion*, can become a potential health hazard. A temperature (thermal) inversion can also be simulated. If the two bottles (one with warm water and the other with cold water) are placed so that the cold water bottle lies on top of the warm water bottle, the conditions in a thermal inversion can be observed. Repeat the experiment, but this time place the warm water bottle on top of the cold water bottle. What happens?

AIR TAKES UP SPACE

Investigation: How can we demonstrate that air takes up space?   (P, I)

Background Information: Air has some unique properties. We cannot see, smell, or taste it, but the characteristics and behavior of air are largely responsible for our weather. Air has no definite shape, but will fill its container. An open paper bag cannot be half full of air because the air will move throughout the bag. Like other gases, air expands and can be compressed. The air over a burning candle expands as it is heated.

All substances occupy space. For air to be considered a substance, it too should occupy space. In the activities that follow, children will demonstrate that air takes up space and therefore is a real substance.

Materials: Plastic food storage bags and twist-ties
A glass or plastic jug
A small piece of modeling clay
A funnel
A glass or plastic tumbler
A basin or tank of water

Procedure: 1. Collect air in a plastic bag and tie the end with a twist-tie. Place the bag on a table top and balance a book on top of the bag. Ask the children what is supporting the book (see Figure 2-5). What substance is between the table top and the book? How does this show that air is real?

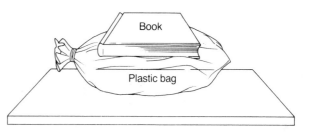

FIGURE 2-5.   What substance is supporting the book?

2. Mold some clay around the neck of a funnel and press the funnel into the mouth of a jug so that an airtight seal is created. (A funnel with its neck in a one-hole stopper may be substituted.) Pour water into the funnel. What happens? How does this demonstrate that the jug was not empty? (See Figure 2-6.)

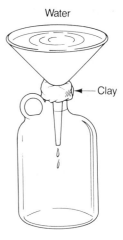

Water

Clay

FIGURE 2-6. What happens when water is poured into the jug through the funnel?

3. Crumple a piece of paper and push it down into the bottom of a clear glass or plastic tumbler. Turn the tumbler upside down and push it directly down into a large basin or tank of water. Pull the tumbler straight out again and inspect the paper. Did it become wet? (See Figure 2-7.) How does this demonstrate that air takes up space?

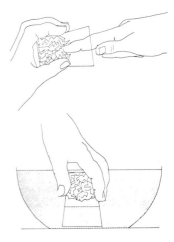

FIGURE 2-7. Why does the crumpled paper remain dry? Reprinted from Kevin Goldstein-Jackson, *Experiments with Everyday Objects* (Englewood Cliffs, N.J.: Spectrum Books/Prentice-Hall, Inc., 1978), p. 5.

Going Further: Interested children may devise other experiments to prove that air takes up space. If air is trapped in a plastic bag and then compressed so that the air enclosed is forced out of the neck of the bag and onto the child's face, further evidence of air's existence is obtained.

Place the end of an eyedropper in a glass of water. Squeeze the bulb and observe the bubbles of air. Why does this demonstrate that air was in the bulb of the dropper?

AIR HAS WEIGHT

Investigation: How can we demonstrate that air has weight?   (P, I)

Background Information: Hard as it may be for some children to believe, air does indeed have weight! This is a characteristic of all substances. The weight of air is due to the gravitational attraction between it and the earth. The air in a basketball or an automobile tire is much denser than the air in the atmosphere because it is compressed in a closed container. A cubic meter of air at standard atmospheric pressure at sea level weighs about 1.2 kilograms (2.64 pounds). The weight of a column of air over a surface of one square meter is about 10,356 kilograms. We do not feel the weight and pressure of air because we are constantly exposed to it and because there are forces and pressures within our bodies which equal the pressure of air upon us. In this activity, children can see for themselves that air does have weight.

Materials: A wire coat hanger
A wire cutter
String
Two balloons of the same size

Procedure: 1. Blow up the two balloons, making them equal in size. Tie the ends.
2. Cut a straight section of the hanger and tie a piece of string around the center so that it is balanced when suspended by the middle string. (Be careful of the sharp edges of the hanger.)
3. Tie a balloon to each end of the hanger wire and adjust the center string so that the balloons are evenly balanced.
4. When the children agree that the balloons are in perfect balance and weigh the same, pop the air out of one of the balloons and observe the effect (see Figure 2-8). How does this demonstrate that air has weight?

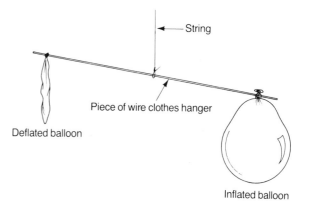

FIGURE 2-8.   Two balloons are balanced on the ends of a wire clothes hanger.

Going Further: Another way to demonstrate that air has weight is to weigh a deflated basketball, soccer ball, or volleyball on a very sensitive balance. Fill the ball with air until it is firm and fully inflated. Weigh the ball again. What causes the difference in weight?

EXPANDING AIR

Investigation: What happens to air when it is heated? (P, I)

Background Information: Air, like other substances, expands as it is heated. If the air in a closed container is heated, the air will expand; but if it cannot escape, this expansion will result in greater pressure exerted on the inside walls of the container. On the other hand, if the walls of this closed container are flexible (as in the case of a balloon), the expanding air will inflate the container when the pressure it exerts on the inside walls is increased.

The expansion of air as it is heated is basically responsible for changes in atmospheric pressure which greatly influence our daily weather. The air in the atmosphere is not in a closed container, so when it is heated, it expands and moves upward. Thus, there will be less air over a heated portion of the earth's surface, and consequently the air pressure is reduced.

Air pressure is measured with a barometer. A falling barometer usually indicates that a warm air mass is moving into the region. Conversely, as air is cooled it contracts, becomes denser, and exerts more pressure on the earth's surface. This condition results in increased barometric pressure, usually signaling the arrival of cool, dry air and generally fair weather conditions. In this activity, children can demonstrate the effects of heating a small air mass.

Materials: An unbreakable glass flask or baby bottle
A balloon
A hot plate

Procedure:
1. Stretch the balloon over the mouth of the flask or bottle.
2. Place the flask or bottle on a warm hot plate (see Figure 2-9). If a hot plate is not available, the flask may be placed over a candle's flame or into a pot of very hot water. What happens? Why does this occur?
3. Using an oven mitt or tongs, remove the flask or bottle from the heat source and place it into a pan of cold water. What happens? Why does this occur? (Because the flask or bottle could possibly break, caution should be exercised in this procedure.)
4. How can this demonstration be related to what happens when an air mass is heated on a larger scale? Remember that in this activity the pressure on the inside walls of the balloon is increased as the air is heated, but in an open system (as, for example, our atmosphere), the air pressure over a given area is actually reduced as the heated parcel of air expands.

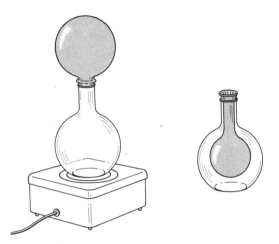

FIGURE 2-9. When air is heated it expands, and the balloon is inflated. When air is cooled it contracts, and the balloon is deflated.

Going Further: The effects of heating air may be observed in other ways. Partially inflate a balloon and measure its circumference. Place the balloon on a sunny windowsill for an hour or two. Measure the circumference again. What happens? What accounts for the difference?

A WATER THERMOMETER

Investigation: How can a water thermometer be made?   (P, I)

Background Information: Most common thermometers operate on the principle that substances expand when they are heated and contract when they are cooled. The majority of thermometers that children use have a column of mercury within a narrow tube. As the temperature rises, the mercury expands and moves upward through the tube. As the temperature declines, the mercury contracts and descends to a lower position in the tube. Water behaves in a similar fashion, and a water thermometer can be made and calibrated which operates on the same principle as a mercury or alcohol thermometer.

Materials: A small soda pop bottle
Modeling clay
A clear plastic drinking straw
Food coloring
An index card

Procedure: 1. Fill the soda pop bottle about four-fifths full of water and add some food coloring.
2. Dry the lip of the bottle and squeeze some clay around the rim. Insert the drinking straw halfway into the bottle and press the clay around the bottle lip and straw to form a tight seal (see Figure 2-10).

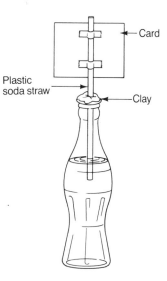

FIGURE 2-10. A water thermometer can be used to demonstrate how liquids respond to changes in temperature.

3. Tape a card to the straw so that a scale can be made. Mark the height of the water in the straw at room temperature. Hold the bottle firmly in your hands to warm it. Does the water in the straw rise? Why does this happen?

4. Place the bottle on a sunny windowsill. Does the height of the water in the straw rise? Why does this happen?

5. Observe and record daily changes in the height of the water in the straw.

Going Further: The water thermometer can be calibrated by marking the card with known temperature readings. Obtain an accurate thermometer and then expose both the commercial and the homemade thermometers to a variety of different temperatures. Mark the height of the water on the homemade thermometer with the actual readings obtained on the commercial thermometer. Estimate the positions for the numbers between the known readings. To obtain a reading for 0° C (32° F), place the water thermometer in a mixture of ice and water. To obtain a reading of 100° C (212° F) place the water thermometer in a pan of boiling water. These two readings may exceed the range of the water thermometer, and the possible reasons for this should be considered.

MAKING A HYGROMETER

Investigation: How can relative humidity be determined? (I)

Background Information: Humidity is an important component of our daily weather. How often we hear, "It's not the heat; it's the humidity!" Our relative comfort on a particular day is usually affected by the amount of water vapor in the air, or the *relative humidity*. If warm temperatures are combined with high humidity, the perspiration from our bodies will not evaporate readily since the air is already carrying a great deal of water vapor, and we feel sticky and generally uncomfortable.

Humidity can be determined in several ways. One way is with a *hair hygrometer*. Human hair stretches when wet and shrinks when it dries. By attaching a pointer to some strands of human hair, calibrations can be made to determine relative humidity. Another, more accurate, means of determining humidity is with a wet- and dry-bulb hygrometer. This instrument is basically an arrangement of two thermometers, one of which has a moist cloth wrapped around its bulb. As water evaporates from the thermometer with the moist cloth wrapped around it, the reading on this thermometer will drop because the process of evaporation uses up heat. (The evaporation of perspiration from our skin has a cooling effect on our bodies.) Using the readings of the two thermometers and referring to a chart called a *humidity table*, the relative humidity is determined. Relative humidity is expressed as a percentage of the amount of moisture that the air is holding compared to the amount it could possibly hold without falling as precipitation. A relative humidity of 70 percent means that the air is holding 70 percent the amount of water vapor that it could hold.

Materials: Two thermometers
A milk container
A cotton shoelace
Rubber bands
A piece of thread

Procedure:
1. Check the thermometers to make sure that they register the same temperature when placed together.
2. Cut a 10 centimeter (4 inch) length of a cotton shoelace and slip it over the bulb of one of the thermometers. Tie it with thread above and below the bulb to hold it in place.
3. Attach the thermometers to adjacent sides of the milk container with rubber bands (see Figure 2-11).
4. Cut a small hole (about one square centimeter) into the milk container, just below the thermometer with the shoelace-covered bulb.

Push the free end of the shoelace through the hole, and fill the inside of the container with water to the level of the hole. This will keep the shoelace wet.

5. The wet bulb may be fanned with a piece of cardboard. As the water around the wet-bulb thermometer evaporates, the temperature is lowered.

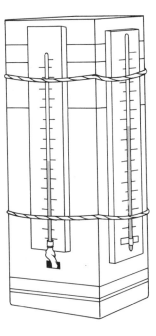

FIGURE 2-11. A hygrometer, an instrument for determining relative humidity.

6. Read the dry-bulb and wet-bult thermometers. Subtract the wet-bulb reading from the dry-bulb reading, and refer this difference to the table of relative humidity (see Figure 2-12). What is the humidity?

Going Further: Some children may enjoy the challenge of making a hair hygrometer. Long strands of hair have to be used and should first be treated to remove oils. The youngsters have to devise a way of fixing the hair between a stationary and a moveable point so that as the hair stretches or shrinks, it moves a pointer over a scale of humidity readings. This apparatus is rather complicated, but with perseverance and interest, children who tackle this unique task will find it quite gratifying.

RELATIVE HUMIDITY (PERCENTAGE) — °C

Temperature of Dry Bulb (°C)	Depression of the Wet Bulb (°C)														
	1	2	3	4	5	6	7	8	9	10	12	14	16	18	20
50	94	89	84	79	74	70	65	61	57	53	46	40	33	28	22
45	94	88	83	78	73	68	63	59	55	51	42	35	28	22	16
40	93	88	82	77	71	65	61	56	52	47	38	31	23	16	10
35	93	87	80	75	68	62	57	52	47	42	33	24	16	8	
30	92	86	78	72	65	59	53	47	41	36	26	16	8		
25	91	84	76	69	61	54	47	41	35	29	17	6			
20	90	81	73	64	56	47	40	32	26	18	5				
15	89	79	68	59	49	39	30	21	12	4					
10	87	75	62	51	38	27	17	5							

FIGURE 2-12. Chart of relative humidity. The depression of the wet bulb is the difference between the wet-bulb and dry-bulb readings.

MAKING A RAIN GAUGE

Investigation: How can the amount of rainfall be measured?   (I)

Background Information: An interesting activity that children may enjoy is setting up a weather station in which they gather information and statistics concerning daily weather. A thermometer is the most obvious piece of equipment to start with, but children may also like to include a hygrometer to measure relative humidity (see pp. 59-61), a barometer, a rain gauge, a wind vane, and other instruments. In this investigation, instructions are provided for the development of an accurate rain gauge.

The simplest rain gauge is a straight-sided can such as a coffee can and a ruler. After a rainfall, the ruler is placed in the can, and the amount of rainfall is easily measured. Often however, the amount of precipitation that falls is so small (1 or 2 centimeters) that it is difficult to accurately measure rainfall in this kind of gauge. For precise measurements of small amounts of precipitation, a system of catching rain in a wide-mouthed container and then pouring it into a smaller, calibrated vessel is more accurate.

Materials: Two straight-sided jars, one with a diameter wider than the other. (A peanut butter jar can serve as the wide-diameter jar, and an olive bottle for the narrow-diameter jar.)
Masking tape

Procedure: 1. Fasten a thin strip of masking tape along the side of the narrow jar. Fill the wide jar with water to a height of exactly 1 centimeter. Pour the water from the wide jar into the narrow jar. Place a mark on the masking tape on the narrow jar that is level with the top of this amount of water. This is the equivalent of 1 centimeter of rainfall as collected in the wide jar. In a similar way, mark the spots on the narrow jar that will correspond to 2 centimeters, 3 centimeters, and other centimeter increments of rainfall. A ruler can be used to mark fractions of centimeters (see Figure 2-13). This procedure will calibrate the narrow jar so that when water collected in the wide jar is poured into it, the amount of water collected can easily be determined.

FIGURE 2-13.   After rain is collected in a wide-mouthed jar, it can be measured by pouring it into a narrow jar that has been calibrated.

2. The wide-mouthed jar should be left outside, away from buildings, to catch rainwater. To prevent tipping, the jar can be placed in a small hole in the ground or braced with soil or clay.

3. After a rainfall, carefully pour the water from the wide jar into the narrow jar. The amount of rainfall can then be determined by sighting the height of the water against the calibrated scale on the narrow jar.

Going Further: Children can record the precipitation that falls during a storm and for longer periods of time such as a week or month by emptying the jar shortly after a rainfall and keeping a running tally of the amount of precipitation during the recording period. Which month of the year had the greatest amount of rainfall? How does this compare with city or county statistics as maintained by the United States Weather Service?

Children may also wish to place rain gauges in different places in their community to see if the same amount of precipitation falls throughout the area.

MAKING A WIND VANE

Investigation: How can wind direction be determined?   (P, I)

Background Information: Wind speed and direction are important indications of weather. Wind direction is determined by observing a wind vane and is stated in terms of the compass bearing from which the wind is blowing. For example, a wind moving from west to east across an area is called a west wind. Wind vanes point in the direction from which the wind is blowing. Knowledge of wind direction can help in locating a low pressure system and the inclement weather usually associated with it. This is why wind vanes are often called weather vanes; they usually provide more information than wind direction alone. Wind speed is measured by an anemometer and indicates, to a certain extent, the differences in air pressure between high-pressure and low-pressure regions.

A wind vane is essentially a delicately balanced arrow with the tail larger than the head. When the wind blows, the larger surface—that is, the tail—is pushed away from the wind, causing the head to point into the wind.

Materials: A pencil with an eraser end
A plastic drinking straw
An index card
A straight pin
A red permanent ink marking pen or red nail polish

Procedure: 1. Make a slit, about 2 1/2 centimeters (1 inch) long, in one end of a plastic drinking straw.

2. Using an index card, cut out an arrow tail and insert it into the slit end of the drinking straw (see Figure 2-14). Some glue may have to be used to keep the tail in place. (If available, a feather may be substituted for the arrow tail and simply inserted into an end of the straw.)

3. With a red marking pen, color the other end of the straw so that it will be conspicuous.

4. The straw should then be balanced on a finger to find its center of gravity. This will not be in the center of the straw, since the side with the tail in it is heavier than the other end. Insert the straight pin through the point where the straw seems to be best balanced.

5. Push the pin into the eraser end of the pencil, and make sure that the straw pivots freely on top of the pencil.

6. This wind vane may then be tied to a longer stick or a stake driven into the ground. Choose a spot where the wind is not blocked by buildings.

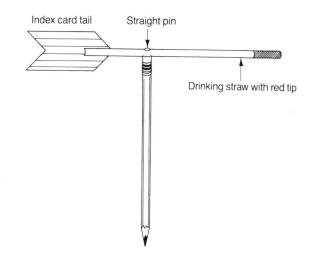

FIGURE 2-14. A simple wind vane can be constructed using a straw, a pencil, a pin, and an index card.

7. On a windy day, look at the direction in which the red end of the wind vane is pointing. Determine the direction of the wind by noting the direction in which the wind vane is pointing and comparing it with the compass reading.

Going Further: Records can be kept of wind direction over a period of time. Does the wind seem to be blowing from one direction more than another? Locations in the middle latitudes are under the influence of the *prevailing westerlies*, a major planetary wind belt. In these latitudes, winds are much more likely to be west winds than in other areas. Subtropical areas are influenced by the *Northeast trade winds* in the Northern Hemisphere, and by the *Southeast trade winds* in the Southern Hemisphere. Do the children's records coincide with these general global wind patterns?

Once youngsters understand how a wind vane operates, they may enjoy constructing variations of the one just described. A carefully balanced, easily pivoting cardboard arrow can also be used to make a wind vane. Many stylized variations are possible.

USING WEATHER MAPS

Investigation: How can weather maps be used as a basis for learning activities?   (I)

Background Information: Weather maps are published daily in newspapers throughout the world. The National Weather Service, a division of the National Oceanic and Atmospheric Administration, is the official meteorological agency. It maintains hundreds of observation stations and keeps weather records. In addition, many stations study the upper atmosphere and temperatures at various levels by means of weather balloons, radiosonde, and other devices. Recently, weather satellites have proven invaluable in determining global weather conditions.

The National Weather Stations are part of an international network of more than fourteen thousand weather stations that are associated with the World Meteorological Organization, an agency of the United Nations. To be useful, the data from these weather stations are assembled on weather maps. Temperature readings usually appear in 10 degree intervals with points of the same temperature connected, forming a line on the map called an *isotherm*. Some maps show *isobars*, which are lines of equal barometeric pressure. Many activities can be developed using weather maps which appear in most local newspapers.

Materials: A weather map from a local newspaper (see Figure 2-15).

Procedure: Using a local weather map, answer the following questions:
1. What was the weather like in your city on the day the map was compiled? What time of day were the readings taken?
2. What was the temperature? Is the barometric pressure indicated?
3. What was the wind speed and direction?
4. How far from you was the nearest rainstorm?
5. Which areas had the highest and lowest temperatures?
6. What is the high temperature likely to be in your city in the afternoon?
7. What is the location of major warm, cold, and occluded fronts? What do these indicate?
8. How is today's weather picture different from yesterday's? What do you think the weather will be like tomorrow?

Going Further: Using weather maps, thermometers, barometers, rain gauges, wind vanes, wet-bulb and dry-bulb thermometers, and cloud study, children can obtain a fairly comprehensive description of existing weather. It may be interesting to have the children "publish" a daily weather report.

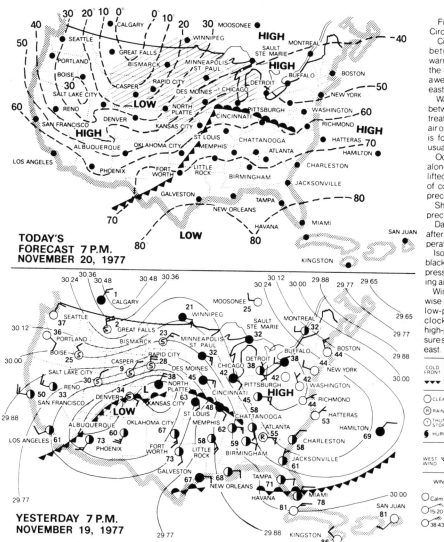

Figure beside Station Circle is temperature.

Cold front: a boundary between cold air and warmer air. under which the colder air pushes like a wedge. usually south and east.

Warm front: a boundary between warm air and a retreating wedge of colder air over which the warm air is forced as it advances. usually north and east.

Occluded front: a line along which warm air was lifted by opposing wedges of cold air. often causing precipitation.

Shaded areas indicate precipitation.

Dash lines show forecast afternoon maximum temperatures.

Isobars are lines (solid black) of equal barometric pressure (in inches). forming air-flow patterns.

Winds are counterclockwise toward the center of low-pressure systems. clockwise outward from high-pressure areas. Pressure systems usually move east.

FIGURE 2-15. Daily weather maps, such as the one reproduced here, can serve as the basis for many interesting activities. © 1977 by The New York Times Company. Reprinted with permission.

Children can also make forecasts by giving special attention to barometric pressure, cloud formations, and weather reports. With these reports and a knowledge of the general movement of weather in their region, children can often achieve a suprisingly high rate of accuracy in their predictions.

3

Energy:
Heat, Light, and Solar

INTRODUCTION

The complex energy problems that currently challenge our society require that today's youngsters develop into intelligent citizens, able to face difficult choices and make important decisions regarding future sources, supplies, and consumption of energy. One way to help equip children to deal with these important responsibilities is to provide them early in their lives with concrete, direct experiences with several forms of energy. In this chapter, energy in its largest sense is explored.

Heat is a form of energy. The sun is the major source of heat in our lives, but heat can also be generated in a variety of other ways. What happens when you rub your hands together rapidly? Can you feel the heat generated? Why do we observe this effect? The nature of heat and its effect upon materials is explored in this chapter.

Light is another form of energy. How different our lives would be without light! For one thing, you could not read this page. Green plants could not grow, for they need light to exist. Could we live without green plants?

Children can probably list hundreds of ways in which light affects their daily lives. The behavior and composition of light are investigated in this chapter.

Solar energy is a matter of great current interest. As solar devices become better understood and perfected, we may see a substantial portion of our energy needs derived from this ever-present source in the years ahead. Some of the activities in this chapter provide children with basic experiences in the principles and effects of solar energy.

As children investigate these distinct yet related energy sources, they will have opportunities to tie their experiences to events that touch them daily. The activities that follow should help children realize the importance of these phenomena and the ways in which we depend on them.

EXPANSION AND CONTRACTION

Investigation: How can the expansion and contraction of solids be demonstrated? (P, I)

Background Information: When materials are heated, they expand; when materials are cooled, they contract. When heat is applied to a substance, its molecules move more quickly, collide with other molecules, and cause these other molecules to move faster. This results in the expansion of the substance because more space is required for the greater movement of the molecules. This is why the liquid in a thermometer rises as it is heated (see pp. 57-58).

Contraction occurs when substances are cooled. The action of the molecules slows down, and the material actually shrinks. Gases also expand and contract as they are heated or cooled (see pp. 55-56). Air masses expand when they are heated, resulting in a less dense air parcel and therefore lowered air pressure.

Many examples of accommodations to these phenomena can be found. Sidewalks are constructed with spaces between the slabs of concrete so that they will have room to expand and contract without cracking. Bridges, railroad tracks, and highways have expansion joints to prevent them from warping or buckling by expansion or contraction.

Water is a notable exception to this general rule that substances expand when heated and contract when cooled. As the temperature of water drops from $4°$ C ($39°$ F) to $0°$ C ($32°$ F) (its freezing point), it *expands* slightly. This unique characteristic is due to the particular arrangement of the atoms in the water molecule and accounts for why the solid, ice, floats on the liquid, water. The expansion of water near the freezing point causes it to be less dense than water that is slightly warmer. Because it is less dense than the surrounding water, it does not sink to the bottom of a lake, but floats on its surface.

In this activity, instructions are provided for the construction of a simple device which clearly demonstrates the effect of heating and cooling on solids.

Materials: A thick wooden dowel or broom handle cut into two 20 centimeter (8 inch) sections

A large-head wood screw

A screw eye (The head of the screw should be just a bit too large to fit through the screw eye.)

A heat source, such as a hot plate, an alcohol lamp, or a cigarette lighter.

Procedure: 1. Screw the screw eye into one end of one of the dowel sticks or broom handle sections. Insert the screw into one end of the other dowel stick (see Figure 3-1).

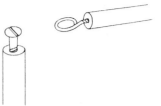

FIGURE 3-1.   Why does heating
the screw eye change its diameter?

2. Demonstrate that the screw head is too large too fit through the screw eye.
3. Heat the screw eye so that it expands sufficiently to allow the screw head to be passed through it.
4. Ask the children why they think this has occurred. Other variations may be tried (such as cooling the screw eye, heating the screw head, and so on), and the effects noted.

Going Further: Children can take a walk through their community and identify modifications that have been made in buildings, walkways, and other structures to prevent cracking or distortion due to the expansion and contraction of solids. Examples include expansion joints or cracks in concrete sidewalks, telephone wires, and railroad tracks.

HEAT CONDUCTION

Investigation: Do all materials conduct heat equally?   (P, I)

Background Information: Heat is transferred from one place to another by three means: *conduction*, *convection*, and *radiation*. In conduction, heat travels through a material. The handle of a pot becomes hot as the pot is warmed because of the conduction of the heat through the pot. This occurs because the molecules in the metal of the pot are caused to vibrate faster as the pot is heated. Molecules in motion collide with adjacent molecules until the molecular motion (and therefore the heat) spreads throughout the entire material.

All materials do not conduct heat equally well. Metals are usually good conductors; wood, hard plastics, and rubber are generally poor conductors. This is why they are often used to make pot handles. A person is not likely to burn his or her hand when using a pot handle covered with a poor conductor, but is quite likely to if the pot's handle is made of metal. Poor conductors are *insulators* and have many domestic and industrial applications (see pp. 93-94). In this activity, children rank materials according to their conductivity of heat. It is both a physical science experience and an intellectual organization task in that the data obtained must be arranged in a meaningful order.·

Materials: A thermometer

A variety of material samples—for example, a piece of wood, a swatch of fabric, paper, metal, plastic, and rubber

Procedure: 1. Have the children touch a variety of materials that have been placed on a sunny windowsill or under a lamp. Ask them to rank the materials from coolest to warmest according to the way they feel. Note the results.

2. Leave the materials on the windowsill (or under the lamp—equidistant from the bulb) for another ten minutes.

3. Then have a child touch the bulb of the thermometer to the surface of each material. Record the results. Are any differences in temperature recorded?

4. Rank the materials again from coolest to warmest. Are these results the same as when the children simply touched the materials with their hands?

5. What conclusions can be drawn about the conductivity of the various materials?

Going Further: Pour boiling water into a teacup or coffee mug. Place a metal spoon in the cup. After a few minutes, ask the children to touch the spoon. Is it hot? Why does this happen? If the spoon actually absorbs some of the water's heat, can placing a spoon in a hot beverage help to cool the beverage?

Ask the children to list the ways in which they use conductors in their daily lives. Once they have had experiences with heat conduction, they will more than likely be able to relate this phenomenon to other events in their environment.

CONVECTION CURRENTS

Investigation: How can convection currents be detected?   (I)

Background Information: Heat energy can be transferred from one place to another by means of currents called *convection currents*. Convection can be noted in the home in the movement of curtains above a radiator when all of the windows are closed. Air, like water, is a fluid and is free to move about. When a parcel of air is heated, its molecules tend to move about quickly and the air expands. Since the total number of molecules in the parcel of air remains the same, the expansion of the substance results in fewer molecules per unit of volume. Therefore, the heated portion is less dense than its surrounding materials, and rises. This occurs in a cup of tea, in ocean currents, and in huge air masses. For example, air above the equator is continually heated and thus rises, causing a convection current on a global scale. Convection currents may be explored readily, but in order to see the movement of air, the air currents must be "tagged." This may be done with a smoking punk, or by igniting a tight roll of slightly moist paper toweling. (For safety reasons, this investigation should be carried out as a demonstration, with an adult holding the source of smoke.)

Materials: A candle

A smoke punk or paper toweling

An incandescent lamp with the lamp shade removed

Procedure: 1. Light a smoke punk. If a punk cannot be obtained, twist a piece of slightly moistened paper toweling into a tight roll, ignite it, and allow it to burn for a while; then blow out the flame. Usually, considerable smoke ensues.

2. Light a candle and hold the smoking punk or paper toweling over it. What happens? The air above the candle is heated and rises, causing the smoke to rise vertically.

3. Hold the source of smoke over an unshaded incandescent bulb that has been lit for a few minutes. What is observed?

4. Heat currents can also be detected by holding long, thin strips of tissue over a heat source. In this case, do not hold the tissue strips over the candle, as this may be a fire hazard.

Going Further: The existence of heat currents can be confirmed by having the children check the temperature at various heights in the room with a thermometer attached to a stick. Have them graph or chart the temperature gradient from floor to ceiling over the same part of the room. Where is the air the coolest? The warmest? How does this demonstrate the effects of convection on the heating of the room? (This investigation is most effective in rooms where the heating devices are on or near the floor.)

HEAT RADIATION

Investigation: How can the effects of heat radiation be observed? (I)

Background Information: Radiation is an important form of heat transfer. The source of almost all of our heat is the sun. This heat comes to us through vast distances of empty space between our planet and the sun. Heat energy radiates outward from the sun in straight lines and in all directions. Similarly, when we sit before a campfire, the front of our body is heated by the radiant energy of the fire, but our back is not heated because the transfer of heat occurs only in straight lines. Heat rays are effective in warming a solid object that obstructs its path (for example, the body in front of the fire) but are far less effective in heating the air because of the low density of matter in it. The amount of heat absorbed by radiation is influenced by the type and color of the material absorbing it (see pp. 86-88).

Materials: A lamp with an incandescent bulb
A thermometer

Procedure: 1. Hold a thermometer about 10 centimeters (4 inches) above a light bulb that has been switched on. Record the thermometer reading.
2. Let the thermometer return to room temperature, and then place it 10 centimeters (4 inches) below the lighted bulb. Again, record the thermometer reading.
3. Is there a difference between the temperature recorded above and below the bulb?
4. If convection can account for the heating of the air above the bulb, what can account for the fact that the temperature reading is higher below the bulb than it is at room temperature?

Going Further: Children can list the ways in which radiant energy affects their lives. Certainly they can feel the warmth of the sun, but have they felt the heat coming from a fire or a radiator? Why indeed are home radiators called radiators?

Radiant heat waves can be focused with a magnifying lens. Outdoors on a sunny day, place a piece of crumpled tissue paper (or dry leaves) on the ground. Placing the magnifying lens between the paper and the sun, try to focus the sun's rays on a small spot. Does the paper catch fire? Why does this occur?

LIGHT TRAVELS IN STRAIGHT LINES

Investigation: How can we demonstrate that light travels in straight lines?   (P, I)

Background Information: Light is a form of energy. It is a form of radiant energy that travels out in all directions, in straight lines, from its source. Other kinds of radiant energy are radio and television waves, ultraviolet rays, infrared rays, microwaves, and X-rays. There is some controversy about the nature of light. Some observations affirm that light behaves as a wave phenomenon, but others indicate that light consists of a beam of particles. It has been suggested that light could be streams of small particles traveling in waves. This is not the kind of explanation that is looked upon with favor in science. A simpler explanation would be preferred! In this demonstration, children will be able to show that light travels in straight lines.

Materials: Three pieces of cardboard
A candle
A hole puncher
Modeling clay
A piece of string

Procedure: 1. Cut three pieces of cardboard of the same dimensions. Punch a hole in the middle of each of the pieces of cardboard in the same spot.
2. Stand the cards up straight in some way. Slotted pieces of wood are fine for this purpose, but the cards may also be set upright by using a piece of modeling clay (see Figure 3-2). Line up the holes in a straight line by passing a string through them and holding it taut. The string should not touch any of the cards, but only pass through the holes.
3. Place a candle on the table top. The wick should be at the same level as the holes in the cards.

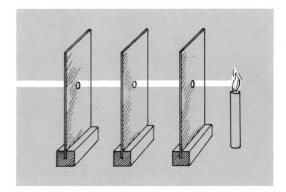

FIGURE 3-2. Light travels in straight lines. Can the candle flame be seen if the position of one of the cards is changed?

4. Light the candle. Can the flame be seen through the holes in the three cards? Now move the cards so that the holes are not perfectly aligned. Can the flame be seen? Why do these results demonstrate that light travels in straight lines?

Going Further: Another demonstration can be used to show that light travels in straight lines. Obtain a length of garden hose or other flexible tubing. Ask the children if they can see light or an object through the tube. In what position is the tube when light can be seen through it? Would it be possible to see light if the tube were not in a straight line?

LIGHT CAN BE BENT

Investigation: How can we show that light can be bent?   (P, I)

Background Information: Light is bent *(refracted)* when it passes from one medium to another. If a stick or pencil is placed in a clear glass of water, it will appear bent or broken (see Figure 3-3). In this case, the light rays pass from one medium (air) to another (water). The change of material caused the light to bend because light does not travel at the same velocity in all media.

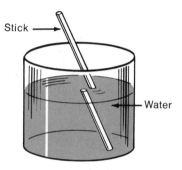

FIGURE 3-3.   The stick appears to be broken because the light rays are bent by the water.

Light travels at a speed of 300,000 kilometers per second in a vacuum. Its velocity in air is about the same, but in a dense medium such as water, its velocity is somewhat less. Perhaps the best way to explain this to children is by the use of an analogy. If the right wheels of an automobile rolling down a highway were to strike soft sand, the automobile would tend to swerve to the right. The velocity of the automobile in sand would be reduced compared to its velocity on the smooth highway. Similarly, a ray of light striking a dense medium such as water will tend to be slowed, and the result is that the whole ray is bent. Conversely, if a light ray enters a less dense medium (air) from a more dense medium (water), the part of the ray entering the air will travel faster than the part of the ray coming from the water. In this activity, the effects of the bending of light can be demonstrated.

Materials: A coin

An opaque cup with straight sides

Procedure: 1. Place a coin on the bottom of the cup and note its position.

2. Slowly add water to the cup. Does the coin appear to move? (See Figure 3-4.) What explanation can be offered for the phenomenon?

3. By carefully positioning the cup and coin with respect to the observer's eye, the coin can be visible when no water is in the cup and then as water is added, seem to disappear. How does this work?

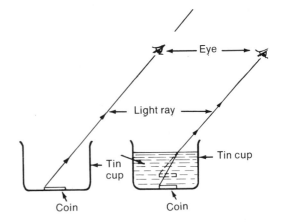

FIGURE 3-4. The coin appears to move because the light rays are bent by the water.

Going Further: Prepare several clear glasses or jars with a pencil or stick in each one as in Figure 3-3, but fill each glass with a different liquid. Possible choices include light syrup, lemon juice, alcohol, vinegar, salt water, and lighter fluid. (Children should not handle any substances that are corrosive or volatile.) Fill each glass or jar to the same level. Observe the pencil in each glass. Does one liquid cause more bending than another? What might account for this? Ask the children to arrange the cups or jars in serial order according to the degree of bending they observe.

THE LAW OF REFLECTION

Investigation: How can the law of reflection be demonstrated?   (I)

Background Information: Light can be reflected by mirrors. When a rubber ball is thrown against a wall at an angle, it bounces off of the wall at about the same angle. When light strikes a flat polished surface such as glass, it reflects or bounces in much the same way that a ball would bounce. If a ball were thrown against a rough surface, we might not be able to predict the angle at which it would bounce back (see Figure 3-5). Clear images are reflected from very smooth surfaces. However, from most surfaces, such as the surface of this page, light is reflected in such a diffuse manner that we cannot see reflected images.

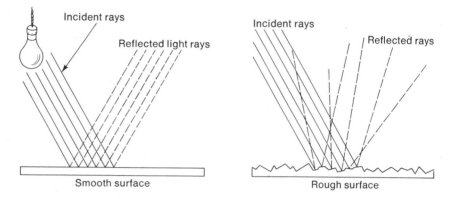

FIGURE 3-5.   Smooth surfaces are good reflectors and rough surfaces are poor reflectors of light.

Physicists have been able to describe the predictable response when light strikes a very smooth surface at a specific angle. This is called the *law of reflection*, and it states that the angle of *incidence* is equal to the angle of *reflection*. Figure 3-6 demonstrates this law. If a broken line is drawn perpendicular to the plane of a mirror, and if two pencils (representing light rays) are aligned in such a way that they are in straight lines with each others' reflections, the angle formed by the light ray approaching the mirror (the *incident* ray) and the broken line is equal to the angle formed by the *reflected* ray and the broken line. (A protractor may be used to confirm the results.) The activity that follows demonstrates this important law.

Materials: A plane mirror
A small hair comb
A flashlight
A piece of cardboard or large index card
Small wads of clay

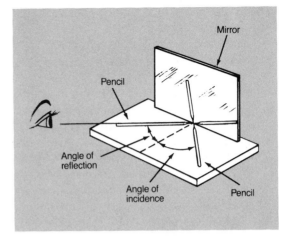

FIGURE 3-6. The angle of incidence equals the angle of reflection.

Procedure: 1. Stand a plane mirror on its long edge on the cardboard. Secure the mirror to the cardboard with wads of clay for support.

2. Hold the comb at a distance from the mirror and shine a flashlight through its teeth onto the mirror (see Figure 3-7). Make sure that the light rays passing through the comb strike the mirror at an angle.

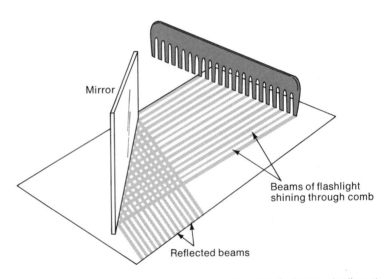

FIGURE 3-7. As the mirror is turned, what changes occur in the angles of the incident and reflected beams?

3. The approaching and reflected beams will shine on the cardboard surface. Do the beams of light seem to strike the mirror and be reflected from it at the same angle? How can this observation be confirmed?

4. Turn the mirror or the comb slightly and notice how the angles change. Do the angles of the incident and reflected rays still seem to be equal?

Going Further: Children may want to list the surfaces on which they have seen a reflection. Some examples might include a lake, a pool of water, store windows, automobiles, metal appliances, and so on. How are these surfaces alike?

MAKING A RAINBOW

Investigation: Of what colors is white light composed?   (P, I)

Background Information: White light is a mixture of many colors. When light is passed through a triangular prism, it spreads out the white light into its component parts, producing a spectrum of colors. Since light may be thought of as a wave phenomenon, when it passes through a prism, this wave moves from one medium (the air) to another (the glass prism). Its component colors, each having a slightly different wavelength (the distance from the crest of one wave to the next), are spread out, or separated by the new medium, the triangular prism (see pp. 79-80).

Color is actually a property of light and not of the object viewed. For example, as white light from the sun bathes a field of grass, light of all colors is shining on the grass, but only the green portion of that light is reflected (see Figure 3-8). All the other colors in the white light are absorbed by the grass. This causes the grass to appear green. Objects that appear white reflect most of the light that shines on them and absorb very little of it. If all the light is absorbed, an object looks black. The color we see is determined by the light that reaches our eyes. The stimuli activate nerve endings in our eyes, and the resulting impulses are then transmitted to the brain, where they are interpreted as color. In this activity, children can produce spectrums from white light and identify the colors the white light is composed of.

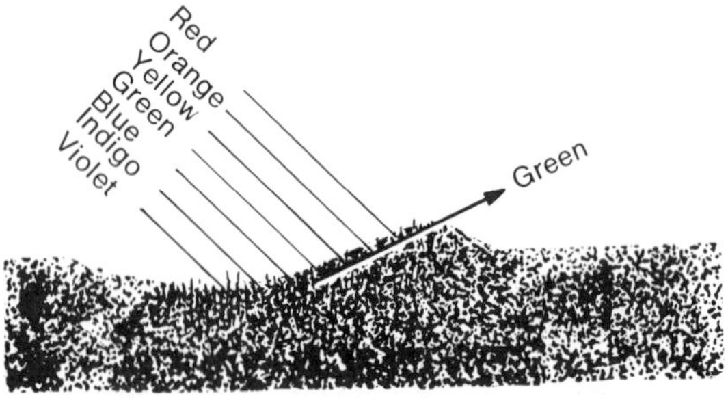

FIGURE 3-8. Grass appears green because it reflects green light.

Materials: A glass triangular prism (If this is not available, a shallow pan of water and a mirror may be substituted. See Step 4 of Procedure.)
A piece of cardboard

Procedure: 1. Cut a thin slit in a piece of cardboard and tape it onto a sunny window.
2. Let the thin beam of light entering through the slit pass through a triangular prism. This should produce a spectrum on one of the walls or

the ceiling of the room. Of what colors is the spectrum composed? Do the colors appear to be in a special order?

3. If a second prism is available, focus the spectrum onto the second prism. The colors should recombine to form white light (see Figure 3-9).

FIGURE 3-9. The colors of the spectrum are separated when white light is passed through a prism. With a second prism, the colors recombine to form white light.

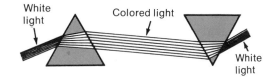

4. If a prism is not available, fill a shallow pan three-fourths full of water. Place the pan near a bright source of light. Place a mirror in one end of the pan (opposite the light source) at a 45 degree angle. Darken the room until the rainbow is visible on a wall or ceiling of the room.

Going Further: Some children may wish to explore other ways of producing a spectrum. Stand a tumbler filled to the top with water on a window ledge in bright sunlight. The glass should project a little over the inside ledge of the window sill. Place a sheet of white paper on the floor and try to focus the spectrum onto it.

Rainbows are formed when sunlight passing through water droplets in the air is reflected by the opposite side of the water droplets and transmitted to our eyes. The water droplets act like tiny prisms when they separate the sunlight into its component colors as the light is reflected to our eyes. To produce an "artificial rainbow," turn a garden hose on and adjust the nozzle until a fine spray results. Stand with your back to the sun and spray the water against a dark background of trees. The rainbow effect should appear near the top of the mist.

HEAT ABSORPTION AND COLOR

Investigation: Which colors of the same material seem to absorb the most heat? (P, I)

Background Information: Heat energy travels out from the sun in straight lines and in all directions. Heat rays warm up solid objects in their path, such as buildings, sidewalks, or people, but they do very little warming of the air that they travel through (see p. 76). Heat waves keep traveling until they are stopped by something solid, and when they do strike an object, the color of the object affects the amount of heat that is absorbed. Light colors reflect (or turn back) much of the heat that strikes them. Dark colors absorb much of the heat. This is why people tend to feel more comfortable wearing light-colored clothing in the summertime. In the activity that follows, children can investigate the relationship between the color of an object or material and the amount of heat absorbed.

Materials: Six different shades of construction paper (Make certain that one is white and one is black.)
Six thermometers

Procedure:
1. On a sunny windowsill or table top, place each of the sheets of construction paper side by side.
2. Insert a thermometer under each sheet of paper. Initially, make sure that all the thermometers have the same reading.
3. If the windowsill is not sunny, shine an incandescent bulb on the papers at an equal distance from all of them.
4. After about an hour of exposure, read the thermometers.
5. Make a chart listing the colors and the temperature readings.
6. Under which color was the temperature the highest? Under which color was the temperature the lowest? What do you think accounts for the differences?

Going Further: Color differences of other materials can be examined. For example, ask children to touch various automobiles on a street or parking lot. (Make sure that all of them are in the sun.) Which colors seem to be the warmest? The coolest? Which automobile color would you prefer in the summer?

The heat absorption of different colors of other types of materials can also be examined. If a variety of colors of the same type of material (for example, cotton quilting, corduroy, metal) can be acquired, place them on a sunny windowsill for an hour. Which colors seem to absorb the most heat? Are the differences in heat more pronounced with materials other than paper?

HEAT ABSORPTION AND MATERIAL

Investigation: Do all materials absorb the sun's heat equally? (P, I)

Background Information: Whenever light or heat falls upon an object, the object absorbs some of the heat. The amount of heat absorbed is affected by the color and type of material exposed to the heat source. The degree of heat absorbed by various substances will influence the types of materials selected for solar collectors, devices used to absorb heat energy from the sun. Naturally, those materials (and colors) which absorb the most heat are most desirable for use in panels and surfaces exposed to the sun when maximum heat absorption is desired.

Materials: Three glass or plastic jars or beakers of the same dimensions
Three thermometers
Dry, light-colored sand to fill one of the jars
Dry, dark-colored soil to fill one of the jars
A heat or projector lamp if a sunny windowsill is not available

Procedure: 1. Fill one jar or beaker about half way with the sand; another with an equal amount of soil; and a third with an equal amount of water.
2. Insert a thermometer well into each jar and make sure that the three reading are initially the same (see Figure 3-10).

FIGURE 3-10. The amount of heat absorbed depends on the type of material exposed to the heat source.

3. Place the three jars on a sunny windowsill, or shine a heat lamp (or a projector light) on the jars and let them heat for 20 minutes.

4. Ask the children to predict what changes will occur in the thermometer readings of each of the jars.

5. After the jars have been exposed to the heat source for 20 minutes, read the thermometers and record the new temperatures.

6. Have children place their hands around each of the jars, and ask them if they can discern notable temperature differences among the three jars.

7. Which jar registered the highest temperature? The lowest?

8. What conclusions can be drawn from the data obtained?

9. What implications do these results have for the use of materials in the design of solar collectors?

Going Further: Other variables can be investigated in similarly designed investigations. For example, does it make any difference if the sand or soil is wet or dry? What differences are registered if clear water, black ink, and red ink are used? Try covering the jars with different materials, for example, aluminum foil, clear plastic wrap, or waxed paper, and note the results in heat absorption. Experiment with the angle at which the sunlight or artificial heat source strikes the materials. Does this angle cause notable differences in the results obtained?

THE GREENHOUSE EFFECT

Investigation: How can heat energy be trapped?   (I)

Background Information: Heat energy can be trapped by glass—as it is in a green-house—in much the same way that it is trapped by the atmosphere. The atmosphere, which is composed of a variety of gases, envelops the earth and traps the sun's energy, keeping the temperature in most places within the fairly narrow range that is supportive of life. When the sun strikes objects on the earth (for example, rocks, soil, water, sand), energy is absorbed by the objects; however, some of the heat is reradiated but cannot pass out of the atmosphere, and thus reradiated heat is trapped within the atmosphere. This investigation demonstrates how a glass "envelope" can trap heat energy, the basic principle upon which passive solar heating systems are designed.

Materials: Two thermometers with the scales etched or printed on them

A large test tube or narrow jar

A one-hole stopper that will fit into the test-tube, and through which one of the thermometers will fit

Procedure: 1. Place one thermometer in the one-hole stopper and then insert the thermometer into one of the test tubes. Place the other thermometer in the open air right next to the one in the test tube. (If a test tube, stopper, and thermometers with etched scale are not available, a jar may be used, and a mounted thermometer may be placed in the jar and the top screwed on with the thermometer inside of the jar.)

2. Place the thermometers in the sunlight or shine a projector light on them (see Figure 3-11).

FIGURE 3-11. What happens when two thermometers are placed in sunlight—one in the open air, one in a glass tube?

3. Have the children read and record the two temperatures every 5 minutes for about 30 minutes.

4. Which thermometer registers the higher temperature?

5. What implications do these results have for the building of sun rooms in homes or apartments? For maximum heating value, which direction should these sun rooms face in the Northern Hemisphere? In the Southern Hemisphere?

Going Further: Discuss some everyday examples of the greenhouse effect with the youngsters; for example, closed automobiles heat up when they are parked in the sun, unobstructed windows permit the sun's warmth to enter an otherwise cool room in winter months, greenhouses retain heat, and so on.

Children can experiment with different sized test-tubes and jars and discern which are the most effective for trapping heat energy. An assortment of glazing materials, that is, clear glass, smoked glass, plastics of various composition and color, can be studied for their relative heat-retaining efficacy.

The angle at which the glazing materials are supported with respect to the sun will also have an effect upon the degree of heat retention. How can this phenomenon be investigated?

A SOLAR COLLECTOR

Investigation: How can a simple solar collector be constructed? (P, I)

Background Information: In the design of any solar collector, the major objective is to trap the sun's energy and convert it into heat. The construction of the simple solar collector outlined in this investigation demonstrates the problems in designing and building these devices for use in space heaters and hot water heaters. The basic principle upon which the collector works is the greenhouse effect. (See previous investigation.)

Materials: A 1-pound coffee can with its plastic lid
A laboratory thermometer with the scale etched or printed on it
Clear plastic wrap or cellophane
Black spray paint

Procedure:
1. Spray the inside of the coffee can with black paint. Why should this help to trap heat? (See p. 86.)
2. With a large nail or metal punch, pierce a hole in the side of the can just large enough for the thermometer to fit through. (The inside of the can will have to be supported with a block of wood or some other object when the hole is punched in order to prevent deformation of the can.) If a laboratory thermometer is not available, use a mounted thermometer that is small enough to be placed entirely inside the coffee can when the can is capped with its lid.
3. Cut out the center of the can's plastic cover, leaving a 1 centimeter (½ inch) rim.
4. Stretch the plastic wrap or cellophane across the open end of the can, and secure it tightly by placing the plastic rim over it and onto the can (see Figure 3-12).
5. Insert the thermometer into the hole on the side of the can. (Place it inside the can if a mounted thermometer is used.)
6. Place the coffee can on a sunny windowsill with the plastic-covered top facing the sun. Prop it up with books so that it does not roll.
7. Record the thermometer reading every other minute for about 15 or 20 minutes.
8. What was the final temperature in the coffee can?
9. What is the room temperature?
10. Is this device effective for trapping heat energy?
11. How could a larger device like this one be used to provide heat for a room or water?

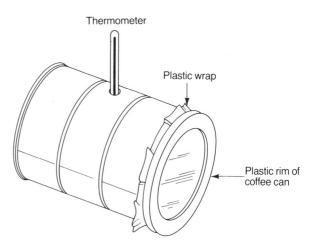

FIGURE 3-12. A simple solar collector can be constructed using a coffee can, black spray paint, plastic wrap, and a thermometer.

Going Further: There are numerous possibilities for variations of this project to investigate further the factors that make for effective solar collectors. Different types of cover materials (glazing) may be explored. For example, is waxed paper, red cellophane, or acetate most effective in producing the highest temperature after 15 minutes of exposure to the sun? Is black the most effective color for the inside of the can? What if the can is unpainted? If the angle or slant of the can with respect to the sun is varied, does any particular position prove more effective than another?

The can may be nested inside of a larger (3-pound) coffee can. (Another hole would have to be punched in the larger can to permit the thermometer to pass through both cans.) In the space between the two cans various insulation materials can be inserted and compared for their effectiveness in retaining the heat collected. Insulation materials might include shredded paper, styrofoam pieces, cotton, rubber pencil erasers, and so on.

THE EFFECTS OF INSULATION

Investigation: How does insulation affect the amount of heat retained in warm water?   (I)

Background Information: A great deal of energy is wasted whenever heating of uninsulated spaces occurs. Insulators are materials that reduce heat transfer from one space to another. In this investigation, children evaluate the relative effectiveness of a variety of insulation materials in terms of their ability to retain heat in a jar of warm water.

Materials: Two identical glass jars with a capacity of approximately 1 liter (1 quart) each

A cardboard box that is larger than the size of one of the jars

Two thermometers

A variety of insulation materials, for example, shredded newspaper, foam rubber, small pieces of styrofoam, fiberglass insulating material, cotton, and so on.

Procedure: 1. Fill the two jars with equal amounts of tap water, and label them A and B.

2. Record the initial temperature of the water in each of jars. (These readings should be approximately the same.) Screw on the jar lids.

3. Place the jars on a sunny windowsill for about one hour to warm the water. Record the final temperature. These readings should be approximately the same. (This step may be eliminated if hot water was used initially.)

4. Cap the jars again and place Jar B deep into a box that has been filled with an insulating material. Place the two jars on a table or desk top that is not in direct sunlight (see Figure 3-13).

5. Record the temperature of the water in each jar after 15 minutes and then again after 30 minutes.

6. Does the box with the insulating materials in it help to keep the water warm during this 30 minute period?

7. Repeat the investigation several times over the course of the next few days, each time using a different insulating material. Fill in a data sheet like the one on the next page for each trial.

8. Which materials seem to retain the greatest amount of heat in the warm water? What implications do these results have for the construction of clothing and homes?

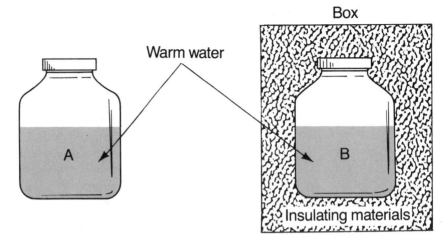

FIGURE 3-13. Two jars of warm water are left to cool on a table top. How effective is the insulating material in retaining heat in the water?

DATA SHEET

Date _____ Insulation Material _____

Thermometer Readings

	Jar A	Jar B
Initial temperature	_____	_____
Warm water	_____	_____
Cooling period after 15 minutes	_____	_____
after 30 minutes	_____	_____

Going Further: The children can be asked to devise other types of insulating techniques—for example, wrapping one of the jars in aluminum foil, covering it with various colors of materials, or placing one of the jars in a closed box without any insulation materials to determine the effectiveness of trapped air as an insulator. The children can experiment with other types of containers, such as metal cups, plastic cups, paper cups, or styrofoam cups.

The effectiveness of various insulation materials in maintaining the temperature of cold drinks can also be explored. Beginning with cold drinks, which materials seem most helpful in maintaining cool temperatures? Are the materials that insulate best against heat loss the same as those which are most effective in keeping drinks cool? What implications can be drawn from these studies for insulating homes or for manufacturing jugs or other drink containers?

TEMPERATURE DIFFERENCES ON A HILL

Investigation: What temperature differences can be found on two sides of a hill? (P, I)

Background Information: Most landscapes are not perfectly flat, but rather they have slopes, folds, hills, and other landforms that can affect an area's climate. Sloping surfaces may face in any direction. On some slopes or hillsides, the surfaces may be situated so that they receive the sun's most direct rays, that is, those rays which strike the surface at a right angle. Other surfaces may be situated in such a way that they rarely receive warmth from the sun. Therefore, the slope or tilt *and* the direction in which hillsides face greatly affect the amount of sunshine they receive. In this activity, children can check the warming effect on two sides of a simulated hill.

Materials: Two thermometers

Two pieces of cardboard, each approximately 25 centimeters by 30 centimeters (10 inches by 12 inches)

Transparent or masking tape

Procedure: 1. Tape the long edge of each of the pieces of cardboard together so as to make a hinge (see Figure 3-14). This will be the "hill."

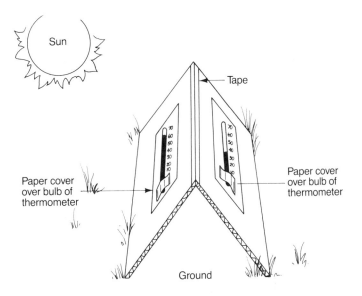

FIGURE 3-14. What is the temperature on each side of the cardboard "hill" after half an hour's exposure to the sun?

2. Tape a thermometer onto each of the two exposed surfaces of the cardboard hill.

3. Tape a piece of paper or index card over the bulbs of the thermometers to prevent them from being exposed directly to the sun.

4. Place the hill on the ground with one side facing the sun directly, and the other side away from the sun and in the shade.

5. Record the thermometer readings when the hill is first put in place and then again after 30 minutes of exposure. How do the temperatures compare? What implications do these results have for home or school construction? In which climates would you prefer to have the windows of a house or school face south? In which climates would you prefer to have the windows face north?

Going Further: Interested pupils can use protractors to measure the angle the slope makes with respect to the ground. Which angle results in the greatest heat on the sunny surface? A compass can be used to determine the precise north-south orientation of the cardboard hill. How do these variables, that is, compass direction and the angle of the slope, affect the heating of the two surfaces? What is the latitude of your community? Is there any relationship between the latitude of your city, the angle of the cardboard hill's maximum heating, and the latitude of the spot where the sun's most direct rays reach the earth?

4

Studying Ourselves

INTRODUCTION

Every child carries around a natural laboratory at all times! The human body provides countless opportunities for youngsters to investigate the wonders of the operation of many of our body parts and systems. As children begin to explore their vital functions, they are helped to understand how we are similar to other members of the animal kingdom and what makes our species unique.

Children can begin their explorations of the human body by taking a look in the mirror. What do they see? What are the most prominent features of the human body? When they look at each other, what do they notice first—hair, eyes, height, clothing? How are we as humans similar and different? How are our bodies adapted for our erect, walking-on-two-feet life style?

In the activities that follow, little equipment is required. Curious minds, willing participants, and live human bodies head our materials lists for these explorations. As the youngsters become involved in these investigations, they will gain firsthand knowledge about their senses, body functions, and some of the intricacies of life itself.

HOW HUMANS ARE SIMILAR AND DIFFERENT

Investigation: How are humans similar and different? (P, I)

Background Information: Humans, all members of the same species *(Homo sapiens)*, are alike in thousands of ways. We all have two arms, two legs, two eyes, two ears, two kidneys, two lungs, a four-chambered heart, and so on. Yet despite all these apparent similarities, we are all different. We exhibit a variety of hair and skin types. Our hair may be straight and black, blond and curly, black and curly, brown and wavy. The children in any group vary in height, eye color, arm length, head circumference, and other characteristics. Young children should first describe their own features, clothing, and characteristics by looking into a mirror, before they begin to draw comparisons with other individuals.

Materials: A mirror
Chart paper and markers
A measuring tape

Procedure: 1. Discuss our membership in the species *Homo sapiens* with the children. Have them list some of the many ways in which we are all alike and how we differ as a group from other animals, for example, apes, dogs, fish, and so on.
2. Identify several human characteristics the children will focus on in their discussion of differences among them. Common features might include height, hair color and texture, head circumference, eye color, arm length, shoe size, and so on. Ask the children to list their own hair color, eye color, and the like on individual worksheets. A mirror may be used for aiding in this task. It might also be helpful for children to work in pairs and fill in the information for one another.
3. Compile the data gathered in charts, graphs, histograms, or other convenient methods for listing the results. What is the most common hair color, hair texture, or eye color? What is the average height, head circumference, and arm length? (See Figure 4-1.)
4. Prepare a report of a particular group's "vital statistics." If more than one section of a grade exists in a school, each class can gather this information and compare the results. If several grades become involved, the youngsters can compare the development of height, arm length, and other measurements, and relate them to age differences.

Going Further: Young children may enjoy exploring their own body characteristics by having a helper trace their bodies as they lie on a large sheet of butcher or wrapping paper. Have the children fill in their own features and clothing. They may color their likenesses with paint, crayons, or markers. If this

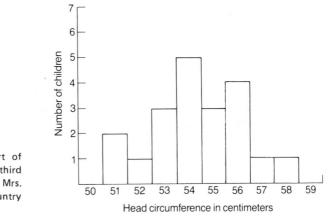

FIGURE 4-1. A bar chart of the head circumferences of third grade boys and girls in Mrs. Smith's class at Riverdale Country School.

activity is accomplished early in the year, it may be repeated later on and the results compared.

A feature that is widely used to identify individuals is fingerprints. Every human being has his or her trademark—no fingerprint is exactly like another. Children can take their own fingerprints by placing a thumb (or other finger) on an inked stamp pad and then pressing it on a clean slip of paper. The children can then compare their fingerprints and, by making careful observations, discuss the features that distinguish one fingerprint from another.

SOUNDS IN THE ENVIRONMENT

Investigation: Can objects be identified by the sounds they make?   (P)

Background Information: The five senses are as fundamental to the young child learning about the world as they are to the scientist studying chemical reactions in a laboratory. Just as the scientist notes color changes, odors, sounds, and textural changes, so too the young child learns by perceiving sights, sounds, odors, and textures in the immediate environment. Careful sensory observation is one of the basic skills of science work, just as knowing the sounds of the letters of the alphabet is one of the basic skills in learning to read. As children refine their ability to perceive sensations in their environments, they enhance their ability to learn. In this activity, children focus on the sounds produced by common objects in their environments.

Materials: An assortment of common objects found in the home and school

Procedure: 1. Ask the children to close their eyes and quietly listen to the sounds they hear. They may hear the sounds of traffic in the street, an airplane overhead, other youngsters in the building, music, pets, and other such familiar sounds.

2. Take a "listening walk" indoors or outdoors. As the youngsters quietly walk around, ask them to note the various sounds they hear. When they are back from the walk, the sounds they heard may be written by each child on a piece of paper and shared with the group, or listed on the chalkboard.

3. Ask the children to close their eyes, or have them all face in one direction. Stand in a place where the children cannot see you, and ask them to identify the objects that you drop or produce sounds with. Some interesting examples might include cutting or tearing paper, fastening two papers with a stapler, dropping a pencil, turning an egg beater, closing a book, dropping a fork, or bouncing a ball. Older children can be expected to write the names of the objects they hear as the sounds are produced. Through continued practice of such activities, you may find that the children's ability to identify objects from the sounds they make improves dramatically, and their general listening skills may become enhanced as well.

Going Further: If a tape recorder is available, it is interesting to tape-record common household sounds and play the tape for the youngsters and have them try to identify the sounds they hear. Some particularly familiar sounds are a person brushing his or her teeth, a vacuum cleaner, a telephone ringing, running water, a doorbell ringing, a child running, water boiling, and so on.

LOCATING SOUNDS

Investigation: How can we locate the direction of sounds we hear?   (P, I)

Background Information: Humans can still hear sounds when the hearing in one ear is impaired; however, unless both ears are functioning, they often find it difficult to determine the direction from which a sound is coming. Each ear perceives a sound with slightly different intensity, and it is this difference that helps a person to tell from what direction a sound comes. A similar situation occurs with eyesight. Although an individual can still see objects with one eye, two eyes are necessary to determine the depth of objects, or where one object is in relation to another when viewed from a distance. In this activity, youngsters will be able to experience the value of two ears in trying to perceive the direction from which a sound is coming.

Materials: Objects to clap together (stones, blocks, or sticks)
A handkerchief or other material to use as a blindfold

Procedure: 1. Place a student in the center of the room and cover his or her eyes with a blindfold.
2. Arrange other children in a circle around the sides of the room. Provide each of them with a pair of objects which, when clapped together, produce a short, sharp sound. (One way is to give each child two stones or blocks.)
3. Ask the child in the center of the room to cover one ear by placing a hand tightly over it.
4. Have one youngster hit his or her stones or blocks together, and ask the blindfolded child to point in the direction of the sound. Repeat this procedure several times, and ask one child to record the number of times the blindfolded child correctly determines the location of the sound.
5. Have the blindfolded child uncover his or her ear and again point in the direction of the sounds. Does the record of correct responses improve?
6. Repeat the procedure several times giving more children a chance to locate the directions of sounds. Are the results always the same?

Going Further: Children may wish to explore other aspects of their hearing. To gain some information on how well we hear, hold a watch near a child's ear. Then move the watch out from the ear until a point is reached at which the sound can no longer be heard. Measure and record this distance. Repeat this procedure for the other ear. Is the distance the same? Have the children use this method to check each other's hearing. If there are children who have more difficulty hearing the watch than others, they should be referred to the school nurse or a physician.

Children may also explore the function of the external ear flap (pinna). Select one child to read from a book. (Playing a phonograph record, if available, will produce even better results.) While the child is reading (or the phonograph playing), have the other children cup their hands behind their ears to create an even larger ear flap. Does the loudness with which they perceive the sound increase? Then ask them to cup their hands in front of their ears with their palms facing the rear of the room. Is the sound they hear louder or softer? What does this demonstrate about the function of our external ear flaps?

SEEING WITH ONE EYE

Investigation: How can the effects on perception of the loss of sight in one eye be demonstrated? (I)

Background Information: Our perception of depth and three-dimensionality is due to the fact that we have two eyes; each eye receives a slightly different scene, and the brain's interpretation of these different scenes allows us to judge distances. Cameras with two lenses (called stereo cameras) essentially do the same thing as our two eyes; they take two pictures, and when the slides are developed and viewed through a stereopticon, there is an illusion of three-dimensionality. In this activity, the children can experience firsthand the effects on perception of the loss of sight in one eye.

Materials: A rubber ball
Material for a blindfold

Procedure: 1. Prepare a blindfold that will completely cover one eye.
2. Have the children play a game of catch in groups of two using a small rubber ball. One child in each pair should blindfold one eye. Have the children in each pair stand about 3 meters (10 feet) apart and toss the ball back and forth, keeping a record of catches and misses.
3. After about 20 or 25 tosses, have the blindfolded child remove the blindfold and continue for another 20 or 25 trials.
4. How does the record of catches and misses when one eye is used compare with the record when both eyes are used? If a difference in results occurs, what might be the cause of it?

Going Further: Children can devise other activities to check on the effects of using one eye instead of two. Is a difference in accuracy noted on work done close to the eyes, such as in completing mazes or dot-to-dot books? Must an object be moving in order to demonstrate the effects of using only one eye? Place two objects that a child has never seen before on a table far away from him or her. The objects should be placed so that one is slightly nearer to the child viewing them. Can a child with one eye blindfolded tell which object is nearer to him or her? What happens if familiar objects are used? If a difference in the child's ability to tell which object is closer occurs when familiar objects are used as opposed to unfamiliar objects, what might account for this?

IDENTIFYING OBJECTS BY THEIR ODOR

Investigation: Can objects be identified by their odor?   (P, I)

Background Information: The sense of smell, essential for the existence of many animals, is not well developed in humans. Odors in our environment, though, often give us clues to what is near or present. Some odors conjure up pleasant images for us—for example, the smell of a homemade apple pie baking in an oven; others tell us that a lavatory has been recently washed with a disinfectant or ammonia, or caution us that something is burning in the kitchen. Our reaction to a particular odor has much to do with our previous experience with the substance. Children are aware of odors in their environment but may never have verbalized the association of a specific object with an experienced odor. In this investigation, children gain experience in identifying objects based upon their odor alone.

Materials: A variety of objects that have a distinctive odor, such as:

cloves or other spices	mint
perfume-soaked cotton balls	fish food
soap	crayons
orange peel	vinegar
coffee	lemon juice
vanilla	moth balls
pencil-sharpener shavings	bananas

Small containers to keep the substances in which allow their odors to escape but prevent the substances from being seen. Small pill bottles, 35 millimeter film cans, or nontransparent salt or pepper shakers are fine for this purpose.

Procedure: 1. Ask the children to walk about the room and stand next to an object that has an odor. Discuss the various objects the children have found.

2. Bring in some of the substances that have odors (see the preceding list), and place them in small containers that permit their odors to escape but do not reveal their contents. (If film cans or pill vials are used, their caps should be pierced with a nail or other sharp object.)

3. Ask several youngsters if they can identify the objects based upon smell alone. If they cannot identify the objects, ask them to describe the odors to the best of their ability. With what do they associate these odors?

4. Discuss odors that the children recognize, like, dislike, associate with danger, and so on.

Going Further: Children can experiment with ways to mask odors. Placing clear plastic wrap around the substance or container may be effective. Try room or air fresheners. How long do their effects last?

Odor "fatigue" can easily be demonstrated. Ask a child to smell some vanilla extract for a minute or two. Soon, the sense of smell will fatigue and the child will temporarily lose the ability to identify new odors. Repeat the activity with other children.

An interesting sorting activity is having children distinguish whether objects have an odor. Provide each participant with a tray of various objects. Ask the children to sort the objects on their trays into two piles—those which have an identifiable odor, and those which do not. Are all the children's findings the same?

To demonstrate how odors are dispersed within a room, have the children sit in various parts of the room so that they are fairly evenly distributed. Stand in one corner of the room and open a bottle of ammonia. Ask the children to raise their hands as soon as they detect the odor. What is the pattern of dispersal? How long did it take for all of the children to experience the odor? Air out the room until the odor completely dissipates.

TASTE AND SMELL ARE RELATED SENSES

Investigation: How are the senses of taste and smell interrelated? (I)

Background Information: The senses of taste and smell are said to be chemical senses. A substance must be in solution or dissolved in saliva to be tasted. For example, if the surface of the tongue is wiped dry, some substances, such as salt or sugar, cannot be tasted. Similarly, materials in the air that are perceived as having odors have to first become dissolved in the fluid that covers the mucous membranes in the nostrils.

These two senses are interrelated. When you hold your nose, many substances seem to have no taste. A blindfolded person who holds his or her nose cannot taste whether he or she is biting into an apple or an onion. Children are usually amazed to find that they have no clues as to what they are tasting if their sense of smell is temporarily cut off.

Materials: Small cubes of raw potato, pear, apple, and onion
Toothpicks

Procedure:
1. Divide the children in the group into working pairs. Give each pair two samples of each of the food cubes. A toothpick may be inserted into each cube.
2. One member of each pair should be blindfolded and should hold his or her nose as the other child presents each food sample, one at a time. The "taster" should try to guess the name of the food he or she has eaten.
3. Can the sample be identified by taste alone?
4. The partners should then switch roles and repeat the activity.
5. Discuss why foods sometimes seem tasteless when we have a heavy cold.

Going Further: Taste buds on the tongue are sensitive to the foods we eat and transmit information about taste to the brain. The four basic taste sensations we perceive are sweet, sour, bitter, and salty. Other tastes are due to various combinations of these four.

Children can try to "map" their tongues by seeing which areas are sensitive to specific tastes. Dip separate toothpicks (or cotton swabs) in salt water, a sugar solution, vinegar (or lemon juice), and black coffee or tea. Touch each toothpick (or swab) to spots all over the tongue, rinsing the mouth after each test. Which areas are sensitive to sweet, sour, salty, and bitter substances? Ask the children to sketch their tongues and mark the areas where they found the strongest response for each taste.

The taste buds for sweet and salty foods are located near the tip of the tongue. The taste buds for sour foods are near the edges of the tongue, while those for bitter food are at the base (rear) of the tongue.

A TEXTURE-MATCHING GAME

Investigation: Can materials be matched by their feel alone? (P)

Background Information: The skin is the major organ of the sense of touch. It is laden with sensitive nerve endings that detect touch, pressure, pain, heat, and cold. Each of these sensations can be stimulated separately.

In this activity, children are asked to match texture samples using only their fingers and not relying upon any other sense. This activity may be developed as a game the children will probably enjoy playing from time to time if it is available to them.

Materials: A cardboard box (larger than a shoe box)

A variety of texture sample pairs (two swatches of the same material), each about 10 centimeters (4 inches) square. Some interesting samples might include corduroy, velvet, sandpaper, aluminum foil (mounted on cardboard), fur, and burlap.

Procedure:
1. Cut a hole in the top of the cardboard box just large enough for a child's hand to pass through, but not so large as to permit the objects within it to be seen.
2. Put one sample of each of the texture pairs in the box.
3. Place the other texture samples in a child's hand one at a time, and ask him or her to reach into the box with the other hand and try to find the same texture. The child should then pull the texture sample out of the box and check visually to see whether a match has been found.
4. Continue until all of the textures have been matched.
5. Interest in this activity can be maintained by changing the materials from time to time.

Going Further: There are many activities involving the sense of touch that can be created or improvised. Some teachers of young children maintain a texture box in the room at all times. Samples of interesting textures are kept in such a box, and children take turns closing their eyes, reaching into the box, and describing the texture they feel. This is a particularly useful activity for aiding in the language development of young children.

Older children may enjoy experimenting with stimulation of the various tactile sensations—that is, with touch, pressure, pain, heat, and cold. The pressure sensation can be felt as a pencil is pressed against the skin. Differences in the heat of objects can be investigated by touching similar objects, some of which have stood in the sun and others which have not. Of course, good safety habits should be developed in such explorations to insure that children will not harm themselves.

INVESTIGATING MUSCLE PAIRS

Investigation: How do various muscle pairs operate? (I)

Background Information: The main movement performed by a muscle is *contraction*. Referring to Figure 4-2, you can see that as the biceps muscle contracts, it pulls the lower arm bones up. This is so because the biceps muscle has two attachments, one at the shoulder and one at the lower arm bones. The triceps muscle relaxes at the same time, loosening the tension on the lower arm so that it may respond to the biceps' contraction. When the arm straightens out, the triceps contracts and the biceps relaxes. Thus, muscles act as opposable pairs. The biceps is a *flexor* muscle; it bends a joint toward the body. The triceps is an *extensor* muscle; it straightens out a joint. Using many pairs of opposable muscles, humans can move their limbs in many directions. Children can investigate the action of several muscle pairs by feeling various contractions as they make certain movements.

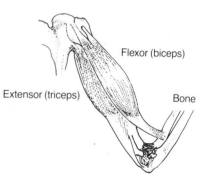

FIGURE 4-2. Flexor and extensor muscles in the arm.

Materials: Chairs and desks or table tops

Procedure: 1. Ask the children to sit down on a chair and place one hand, palm up, inside their desk or under a table top and push *up* on the desk or table top. With their other hand, have them feel the front and back of their upper arm. Is the flexor (biceps) or the extensor (triceps) contracted (hard)? The muscle that feels hard is contracted; the muscle that feels soft is relaxed.

2. Now ask the children to press *down* on the desk or table top. As they touch the two upper arm muscles, ask them to discern which one is hard now. Which muscle's contraction is responsible for the straightening of the arm? How does this demonstrate the operation of opposable pairs of muscles?

3. While the children are sitting in chairs, have them alternately slide their feet back and forth on the floor. With their hands on the under-

sides of their thighs, ask them if they can feel any muscle contraction as they slowly slide their feet.

4. Then ask the children to place their hands on top of their thighs and feel the muscle movements as they slide their feet. Which muscle (upper or lower thigh) contracts as the foot is moved forward? Which muscle contracts as the foot moves back?

Going Further: Other muscle pairs may be similarly investigated. Ask the children to locate other muscle pairs and analyze their movements. Actions of the arms, fingers, tongue, eyelids, and wrists can be explored.

EXPLORING NERVE REFLEXES

Investigation: How can we investigate nerve reflexes?   (I)

Background Information: The nervous system keeps all parts of the body working smoothly together. It coordinates actions and responses so that the organism can function as a whole. Some nerve responses are quite complicated and require thought, action, and memory. Some examples of these complex nerve actions are sewing, cooking, and gardening. All require deliberate, thought-out actions. The simplest form of nervous system control is called the *reflex action.* Some examples of reflexes are coughing, sneezing, blinking, pulling one's hand away from a hot object, and the familiar knee-jerk reflex. Reflexes are inborn—that is, they are present from birth. They are automatic and do not require any thought, and they serve to protect the body from environmental dangers.

There are a variety of reflex actions that can easily be demonstrated. These include the *pupillary reflex* (the opening and closing of the pupils of the eyes in response to altered lighting conditions), the *blinking reflex,* and the *knee-jerk reflex.*

Materials: A sheet or pane of stiff clear plastic or plexiglass

Procedure: 1. Have the pupils work in pairs for all of these reflex demonstrations, with one child serving as the observer. Roles should then be reversed.

2. Ask one child in each pair to cover his or her eyes for at least one minute. As he or she uncovers the eyes, have the other child (the observer) look directly at the first child's pupils (the dark circles within the colored portion of the eyes). What happens? This observation can be even more impressive if a flashlight is shined into the eyes of the child who just uncovered them. Why do the children think this reflex occurs? What value does it have for us? This is called the pupillary reflex.

3. To demonstrate the blinking reflex, have one child hold a pane of clear plastic or plexiglass in front of his or her face while the other child throws a piece of crumpled paper against the plastic, aiming at the eyes. What is the reaction? How is this reflex helpful to us?

4. To demonstrate the knee-jerk reflex, have one child cross his or her legs while the other child lightly chops the first child's crossed leg just below the kneecap with an open hand. What happens to the leg? This method is used by many physicians to check on the operation and integrity of the nervous system.

5. To obtain a sense of reaction time, ask the children to hold hands while sitting in a circle and to pass a hand squeeze from one child to the next around the circle. How long did it take until the hand

squeeze was transmitted back to the first child? If this time is divided by the number of youngsters in the circle, the result is the average reaction time.

Going Further: Children can read about other reflex actions and investigate their operation. Some youngsters might want to diagram the nerve and muscular reactions in a simple reflex action. This information is readily available in encyclopedias and physiology books. Particularly motivated youngsters might check to see if they can interview their own pediatrician or family doctor to gain more information about reflex actions and their importance.

MEASURING LUNG CAPACITY

Investigation: What is the volume of air that our lungs can displace?   (I)

Background Information: Breathing is the process of drawing air into the lungs (inspiration) and expelling it (expiration). It is inside of the lungs that oxygen is supplied to red blood cells and carbon dioxide is removed. Children may be interested in determining the actual volume of air that their lungs can hold at any given time. (Actually only a portion of this total capacity is used in normal breathing.) In this activity, children can calculate their lung capacity by measuring the quantity of water that is bubbled out of a jar when they exhale through a tube into it. Many interesting questions can be answered once this data is obtained.

Materials: A large basin, plastic pail, or fish tank (A sink with a stopper that works well can be substituted.)

A gallon (or 3 or 4 liter) plastic jug or container

One meter (about 3 feet) of plastic or rubber tubing

A vessel with volume markings on it (A large measuring cup, graduated cylinder, or liter container with appropriate markings will be fine for this purpose.)

Procedure: 1. Fill the gallon (or 3 or 4 liter) container with water.

2. With a hand covering its opening, carefully invert it into the larger vessel (also filled with water) so as not to permit any air to bubble into it.

3. Have one child hold the smaller vessel as the rubber tube is placed through its neck (see Figure 4-3).

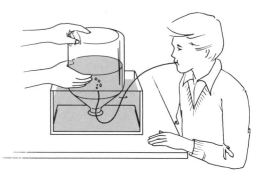

FIGURE 4-3.   A setup for measuring lung capacity.

4. Select another child whose lung capacity will be measured. Ask him or her to take a deep breath and then exhale slowly through the tube. That child should exhale as much air as he or she can but should not exhale to the point of feeling sick or dizzy.

5. With the jug still in the larger vessel or sink, remove the rubber tube, and place a palm firmly over the mouth of the jug and remove it. Be careful not to allow any additional water to escape.

6. Turn the jug right-side up. Using a measuring cup or graduated cylinder, fill the jug to the top again, and record the amount of water required to refill it. This is approximately equal to the volume of air that the child exhaled. The result should be expressed in cubic centimeters, centiliters, or ounces, according to the units on the measuring device.

7. For purposes of comparison, a line can be drawn on the outside of the jug to mark the water level before it is refilled, and the name of the child whose lung capacity the line corresponds with can be written next to it. Which child in the group has the largest lung capacity? Does there seem to be any relationship between the height or weight of the children and their lung capacity?

Going Further: Children can use the data that they collected in the investigation to compute interesting statistics. For example, what is the volume of air in the room that the children are occupying? (This can be found by multiplying the length, width, and height of the room. Be mindful of using consistent units, such as inches, feet, yards, centimeters, or meters.) How many of the average child's lung capacities would fit into the room? (Some long division will have to be used to solve this problem!)

How often do the children breathe in and out in a minute? Does the volume of air inhaled and exhaled in normal breathing come close to the lung capacity as calculated? How can this be checked?

EXERCISE AND THE BREATHING RATE

Investigation: How does exercise affect the breathing rate? (I)

Background Information: When people exercise, they generally breathe at a faster rate than when they are resting, because many body parts (and particularly the muscles) require relatively large amounts of energy in a brief period of time. In order to provide this quick energy, the body consumes a greater amount of oxygen than is normally needed, and so the breathing rate increases. This increased breathing rate not only supplies the body with more oxygen, but also releases greater quantities of carbon dioxide than is usually the case. Children can readily investigate this basic bodily process.

Materials: None

Procedure:
1. Have the children count and record the number of times they breathe in a minute while sitting down.
2. Have the children jump up and down in place 25 times.
3. Immediately after this exercise, have them count and record their breathing rates again.
4. How do the two rates compare? To what do the children attribute the change in breathing rate?
5. How long does it take until the breathing rate returns to the initial, resting rate?

Going Further: The pulse rate is also affected by exercise, and some children may wish to investigate the changes in pulse rate as a result of exercise. First, the children should learn to take their own pulse by lightly applying their index and middle fingers to an area on the underside of the wrist, 2 or 3 centimeters (1 to 1½ inches) below the base of the hand. After locating this pulsation, have the children count the number of times they feel it in 30 seconds and then multiply this number by 2 in order to derive the number of beats per minute. Ask the children to locate other areas of pulsation—for example, in the temple or in the neck near the hinge of the jaw. Is there a wide variation in the pulse rate among the individuals in the group? Is there any difference in the pulse rates of children of different sizes? Between boys and girls? Between children and adults?

Have the children exercise for one minute and compare their pulse rates before and after exercise. What accounts for the differences found? Why do the children think joggers often take their pulse after they have been running for a while? What are the advantages of trying to slow down the pulse rate in terms of cardiovascular efficiency?

FOOD TESTS

Investigation: How can food tests be conducted? (I)

Background Information: Most foods can be categorized as belonging to one of three main groups of substances: proteins, carbohydrates, and fats. Proteins are the basic food materials used for bodily growth and repair; milk, meat, poultry, fish, and eggs are major sources of protein. Carbohydrates are foods that contain some form of sugar or starch. Carbohydrates are the major fuel for our bodies in that they supply most of the calories in our diets. They are also important for quick energy supply. Among the foods rich in carbohydrates are bread, cake, macaroni products, potatoes, and fruits. Fats are another good source of calories; butter, margarine, lard, and vegetable oils are common sources of fat.

As children learn more about their own nutrition and study their dietary habits, they may be interested in conducting simple tests to verify the presence of carbohydrates, proteins, or fats in the foods they eat. Such tests use indicators to detect the presence of the substances under study. In the following activities, simple tests for the presence of starch and the presence of fats and oils are described. (The test for protein involves the use of corrosive acids and other volatile chemicals, and is therefore not recommended for elementary-school-age children.)

Materials: A brown paper bag

Iodine solution

Fatty foods—for example, peanuts, butter, or meat

Foods that contain starch--for example, bread, crackers, halved lima beans, or boiled rice; and some foods that do not—for example, celery or onions.

Procedure: 1. Ask the children to collect some food substances to be tested for the presence of fat or starch.

2. Have the youngsters predict whether or not they think each of the foods collected contains fat or starch. List the predictions.

3. To test for the presence of fat, rub some of the food on a piece of a brown paper bag. Hold the test spots up to the light. If the spot becomes translucent, it indicates that fats or oils are present.

4. To test for the presence of starch, place a drop of iodine solution on the food in question. (Some foods may have to be sliced first to expose the main food substance—for example, slicing a potato.) If dark blue-black stains appear where the iodine was placed, starch is present.)

5. Have the children list the results and check them against their predictions.

Going Further: Particularly motivated children may enjoy conducting other chemical indicator tests. Some that may be familiar to them are testing the acidity or alkalinity of a substance by the use of litmus or pH paper (available in many drug stores), and testing the amount of humidity in the air by using the chemical cobalt chloride (available in hobby shops where chemistry sets are sold, or science supply houses). To test for humidity with cobalt chloride, soak a piece of blotter paper in a solution of cobalt chloride mixed with water. After the soaked blotter paper is removed from the solution, it can be cut and mounted on a child's art work, for example, as the roof of a house, the tongue of a dog, or the hair on a face. If the area covered with the cobalt chloride-soaked paper turns blue, it is generally dry in the room; if it turns pink, it is moist in the room. Usually dry air is associated with fair weather, while moist air is associated with precipitation.

ENJOYING A VARIETY OF FOODS

Investigation: What foods are eaten by the children in a group?   (P, I)

Background Information: In exploring the operation of the human body, children ought to pay some attention to those attitudes, habits, and practices which help to keep the body functioning properly. Rest, exercise, safety habits, and proper nutrition are all keys to good health. A good diet in childhood is essential for optimum growth and development. Food is both the fuel and the building material of the body, and discussing the foods we eat and good nutritional habits is well worth the time spent doing so. A well-planned, balanced diet will include daily intake of items from each of the four basic food groups: meat (including beef, veal, pork, poultry, fish, and eggs); vegetables and fruit; milk and milk products; and bread and cereals.

One of the most effective ways of improving diets is to help children become aware of the foods they eat as well as some they avoid. Surveys of children's eating habits may reveal important information which, along with nutritional guidance, may point the way toward the development of sound eating habits. In this activity, children are asked to keep records of the foods they eat over a period of time. If one or more of the four basic food groups is not represented in a child's diet, then he or she can be helped to analyze dietary deficiencies and find ways to improve overall nutrition.

Materials: None

Procedure: 1. Discuss the importance of good nutrition with the group of young-sters. Explain that through surveying what they regularly eat, we can analyze trends in their diets and point the way toward positive nutritional practices.

2. Ask the children to keep a record of everything they eat for a period of at least three days, one of which should include a weekend day. (Younger children may require the help of their parents in keeping these records, or may be interviewed by an adult to obtain the required information.) All of the foods eaten at each meal, as well as all snacks, should be listed. The importance of being honest in these records is essential, and this ought to be impressed upon the children. Youngsters are apt to forget some between-meal snacks, so carrying a small pad in their pockets during the three-day survey may be helpful.

3. At the end of the three-day period, the records should be discussed with the children. Did their diets include a daily selection from each of the four basic food groups (milk, meat, vegetables and fruits, and bread and cereals)? In which areas are the children's diets adequate? In which areas, if any, are they deficient?

Going Further: Some children may wish to become very specific about their eating records and list foods eaten as well as the quantities consumed. Common measures can be reviewed, such as liter (or quart), a tablespoon, a teaspoon, a slice, and so on. Lists of the recommended quantities of the basic four food groups are available from dieticians, school lunch directors, nutritional councils, and federal agencies.

Children often neglect certain areas of the basic four due to individual preferences, poor eating habits, or food prejudices. The children in a group may wish to list the three foods they like the most and the three foods they like the least. This information can be compared. Are children's favorite foods similar in the group? Are one child's least favorite foods included among the list of another child's favorite foods?

Children can be encouraged to taste new and unusual foods by setting up snack bars with foods planned and prepared by the youngsters, keeping in mind the importance of selecting items from the basic four food groups. If the group includes members of various ethnic and national origins, the children may enjoy preparing native dishes or bringing them from home for others in the group to eat. Such sharing may help the children develop a positive attitude toward trying new and unusual foods and food preparations.

5

The Earth

INTRODUCTION

The earth is the planet on which we live. An important part of every child's education is learning more about our earth home. These activities emphasize the study of soil, rocks, and earth features found in the local area.

Maps of the local area will be useful. Road maps can help. Community and regional planning agencies and soil conservation offices often have specialized maps that can be very helpful. The local topographical quadrangles ("topo maps") can be especially helpful. Write to the United States Geological Survey, Washington, D.C., 20242, for an index map of your state. From the index map you can determine which quadrangles will be helpful. Begin to mark on the maps places that may be of geological interest and where it might be useful and interesting to take children on field trips.

Often there are people and organizations in the community that have a special interest in the study of certain aspects of the earth and who have specialized knowledge. Usually it is possible to arrange for the children to meet with such people.

ALIKE AND DIFFERENT

Investigation: How are the pebbles we find alike? How are they different? (P)

Background Information: One of the important intellectual skills that young children can develop through science activities is finding ways that objects are alike and different. To do this, they examine objects and describe their physical properties. Then they determine which objects have these physical properties and which do not.

Materials: Pebbles from nearby yard and grounds

Procedure:
1. Have the children collect pebbles from nearby yard and grounds.
2. Have each child describe a pebble. Emphasize that they should describe the pebbles in terms of such physical attributes as color, smoothness, and shape. Keep a record of the attributes that they use in their description.
3. Make certain that each child has a pebble. Mention each attribute that has been used. Ask each child to examine his or her pebble to see whether it has this attribute.

Going Further: Have the children group together all pebbles that have a particular attribute and those that do not. Repeat for other attributes. With older children a classification scheme can be devised based on the attributes that have been described. Repeat with other kinds of objects, such as leaves.

SOIL

Investigation: What is our soil like?   (P, I)

Background Information: Most soils are composed of a variety of particles such as sand, clay, and organic materials. By spreading samples of soil out on pieces of paper and examining them, we can learn a great deal about the soil and how it was formed. Magnifying glasses are helpful.

Materials: Samples of soil
Sheets of paper
Magnifying glasses

Procedure:
1. Spread samples of soil out on sheets of paper and examine them for such properties as those in items 2 through 7.
2. *Color.* Dark color usually indicates organic matter. It may also indicate a poorly drained soil. Light colored soils are usually well drained.
3. *Size of particles.* Are the particles larger or smaller than a grain of sand? Clay and silt are made up of very small particles.
4. *Shape of edges.* The shape of the edges indicate the extent to which particles have been worn down by various forms of erosion. If possible, obtain samples of sand from a sea or lake shore. These particles will often have been worn smooth by the pounding waves. Compare with grains freshly broken from a rock, which are usually jagged.
5. *Differences in particles.* How many different kinds of particles are there? What are the differences?
6. *Organic matter.* Is there evidence of organic matter? If so, what might be its source?
7. *Living things.* Are there living things in the soil? What are they?

Going Further: Try to find out how soils are formed. Many soils are formed from the underlying bedrock. What are the processes that help change rock to soil?

What forces can move soil from one place to another? What happens to soil when it is moved by wind, running water, or ice?

MAPS

Investigation: How can we make a map?   (P, I)

Background Information: A map is a representation of an area, such as a community, region, state, nation, or the world. One of the best ways to help children learn more about maps—how they are constructed and used—is to have them construct a map of their own. In this activity, they construct a map of a room, such as a classroom or a room in their home (see Figure 5-1). The children learn how to use a scale and how to make representations of an area on paper.

Most maps are drawn to *scale*. A small distance on the map represents a much longer distance in the area being mapped. An important skill in map making is choosing the most appropriate scale. In using maps, it is important to be able to interpret scales in order to gain a better comprehension of distance.

Materials: Meter stick or yard stick
Protractor
Paper
Pencil

Procedure: 1. Measure the sides of the room to be mapped. With a protractor, try to determine the angles at which the walls meet. One way to do this is to fold or cut a piece of paper so that it fits into the angle made by the walls and then to measure the angle of the paper with a protractor.
2. Select a scale to be used in drawing the map. A convenient scale is having one centimeter on the map represent one meter in the room.
3. Draw the outline map of the room using the scale that has been agreed upon.

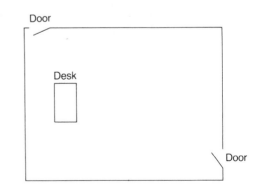

FIGURE 5-1. A ten-year-old's outline map of her classroom. Scale is 1 centimeter equals 1 meter.

4. With the measuring stick and protractor, determine the positions of objects in the room relative to the nearest wall. Locate these objects in the corresponding position on the map.

5. Measure the distance diagonally from one corner of the room on the map to the other. Using the map scale, determine what this distance should be in the room. Measure the diagonal in the room. How close was the estimate from the map to the distance measured?

Going Further: To give children experience in going from the map to the area represented by the map, point to a spot on the map and ask them to state what is located on a corresponding spot in the room.

Give children experience in using such maps as local area or street maps, road maps, and local topographical map quadrangles. They can be asked to locate features on the maps, to give directions from one point to another, and to use the maps to find certain features.

Have the children make a map of the room using a different scale.

SEDIMENTS

Investigation: How do sediments settle under water?   (P, I)

Background Information: Some of the particles that make up soil separate when they are shaken in water and allowed to settle. Under lakes, ponds, and oceans, sediments often settle out in layers and eventually form layered sedimentary rocks. It is possible to learn something about conditions on the land and in the sea by studying sedimentary rocks that were formed under the water.

Materials: Glass jar
Samples of soil

Procedure: 1. Fill a glass jar about three-fourths full of water.
2. Pour a sample of soil into the water.
3. Cap the jar and shake the soil and water mixture. Allow the mixture to stand for some time, then examine and make observations such as those in items 4 through 6.

Organic material

Suspended material

Clay and silt

Fine sand

Coarse particles

FIGURE 5-2. Shake soil in water; then allow the soil to settle. Are layers formed? Do some materials float at the top?

4. *Layers.* Does the soil settle to the bottom in layers? If so, what are the differences between the layers? What kind of material settled out first? last?
5. *Floating material.* Is there material floating on top of the water? What kind of material does this seem to be? (Often this is organic material.)
6. *Suspended material.* What is the color of the water? What kind of material seems to be suspended in the water? Allow the water to stand for several days. Does all the material eventually settle out of the water?

Going Further: Obtain samples of sedimentary rock. Are there layers in the sedimentary rock? If so, are there differences between the particles in the layers? Try to imagine the conditions under which the particles that make up the rock may have been washed off the land and deposited under the sea.

SOIL PROFILE

Investigation: What kind of soil do we have underground?   (P, I)

Background Information: For farmers, gardeners, and home builders it can be important to know what kind of soil we have underground. A cross section of the
soil underground is called the *soil profile*. We can see the soil profile in a
road cut, stream bank, or in the excavation for a building. Sometimes
gardeners dig a hole in a garden, keeping one side vertical, in order to examine the profile. An examination of the soil profile can tell us a great deal
about how the soil was formed, what has happened to it, and how it can be
used.

Materials: Meter stick

Spade

A magnifying glass if you wish to examine soil particles

FIGURE 5-3. A soil profile
showing the topsoil and subsoil.
Can you find the bedrock?

Procedure: 1. Find a place, such as a road cut, stream bank, or recent excavation,
where the soil profile is exposed. In some cases it may be necessary
to dig a hole to expose the soil underneath the surface.

2. *Layers*. Most soils have fairly definite layers. Usually, the *topsoil* near the surface is relatively dark, while the deeper *subsoil* usually has a lighter color. If there are no layers, the soil has probably been moved about and mixed up. This often happens where new homes or schools are built and there is considerable landscaping.

3. *Depth*. Beneath the soil, there is usually *bedrock*. How deep is the soil? In some cases, the bedrock may be so deep that it is not exposed. Deep soils are often relatively old soils; shallow soils may indicate erosion.

4. *Thickness of layers*. Measure the thickness of the layers. A thick topsoil layer usually indicates fertile soil.

5. *Color*. Dark brown or black colors usually indicate the presence of organic matter. Red or yellow colors usually indicate good drainage. If the color of the subsoil is similar to that of the bedrock, it was probably formed from the bedrock.

6. *Feel*. Squeeze a moist sample of the soil between the fingers. If the soil feels gritty, it probably contains a great deal of sand. If it can be squeezed into a smooth smear, it probably contains a great deal of clay. Many good soils contain both sand and clay.

7. *Pebbles and rocks*. If pebbles and rocks are spread throughout the soil, the soil material was probably deposited by a glacier. Many of the soils in the northern regions of North America, Europe, and Asia were deposited by glaciers.

Going Further: Examine soil profiles in other places and regions, and compare them with those near your home or school.

WATER IN SOIL

Investigation: How much water can different kinds of soil hold? (P, I)

Background Information: A soil's ability to hold water is an important characteristic. If a soil can hold a great deal of water, plants that grow in the soil can survive extended periods. On the other hand, in constructing a baseball field, it is desirable to have soils that do not hold a great deal of water. In this activity, children can compare different soils in terms of their ability to hold water.

Materials: Different soils, such as sand, clay, and ordinary garden soil

Containers open at both ends (Lamp chimneys work well. Coffee or juice cans that have had both ends removed can also be used.)

Pieces of cloth that can be fastened across one end of each of the containers.

String or rubber bands

Graduates or measuring cups to measure soil and water

Procedure:
1. Obtain samples of different kinds of soil, including sand, clay, and soil from a local field or garden.
2. Fasten pieces of cloth across one end of the open containers with string or a rubber band.
3. Support each of the containers so that the cloth-covered ends are at the bottom and so that water can easily flow through. One way to do this is to support them on blocks or pebbles. Pans or jars should be placed under the containers to catch the water that flows through the soils (see Figure 5-4).
4. Measure or weigh equal amounts of each of the different kinds of soil.

FIGURE 5-4. Pour equal amounts of water into equal amounts of soil. How much water will each soil hold?

5. Put each soil sample into a separate container. Shake the containers so that the soil will settle uniformly.
6. Measure out equal volumes of water for each of the soil samples. Some experimenters have measured out volumes of water equal to the volumes of soil.
7. Pour the equal volumes of water into each of the containers.
8. Measure the amount of water that flowed through each of the soil samples.
9. Subtract the amount of water that has flowed through the soil from the total amount that was poured onto the soil. The remainder is the amount of water held by the soil. Which soil held the most water? The least?

Going Further: Examine the color of the samples of water that have passed through the different soils. In which case does there appear to be the most material leached out of the soil? Evaporate equal amounts of water from each of the samples. In which case is there the greater amount of residue? Which kind of soil would you prefer for a garden? For a baseball field?

WATER MOVES IN SOIL

Investigation: How does water move up and down in soil?   (P, I)

Background Information: Water travels downward though soils by a process called *percolation*. This is most likely to occur in areas where there is considerable rainfall. As the water percolates downward, water-soluble materials will dissolve and be leached away. Since alkaline materials, such as various salts, tend to be more soluble than acidic materials, the less soluble materials tend to remain, and the soils become acidic.

Water travels upward through soils by a process called *capillary action*. This is most likely to occur in areas such as deserts, where there is relatively little rainfall. Here, too, alkaline materials dissolve in the water and are carried to the surface of the soil. When the water evaporates, the alkaline materials are left as residue. Often desert soils are covered with gray or white alkaline materials such as salt.

Materials: Two lamp chimneys or other containers with open ends, such as juice or coffee cans
Soil
Two pieces of cloth
String or rubber bands
Water
Pans or other containers for water

Procedure: 1. With string or rubber bands, fasten the pieces of cloth over the small ends of each of the lamp chimneys or over one end of the juice or coffee cans.

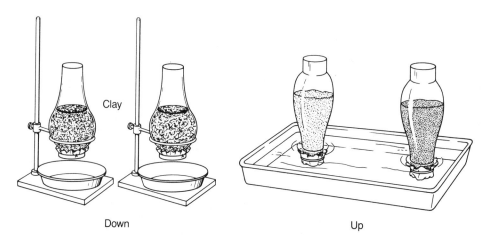

Down

Up

FIGURE 5-5.   Water can move both up and down in soils.

2. Fill each of the lamp chimneys or containers about two-thirds full of soil.

3. Set one of the lamp chimneys or containers in a jar so that the water that runs through the soil will be collected in the jar.

4. Pour a given amount of water into the lamp chimney. With soils that are not too porous, it may be convenient to fill the lamp chimney with water. Watch the water move down through the soil. How long does it take before water runs out of the soil down into the container? If none does, more water should be added. Different soils may be tried to see which allow the fastest percolation.

5. Take a little of the water that has percolated down through the soil and evaporate it. Is there any residue left?

6. Place the other cloth-covered lamp chimney in a pan of water. (Plant trays that are used for growing plants work very well.) Observe to see if water moves up through the soil. If it does, how long does it take for the soil to become moist at the top? Different soils may be tried to see which allow water to move upward the fastest.

Going Further: Put salt water into the tray and maintain the level of water so that the water will continue to move up through the soil for a number of days. Do you eventually see a trace of salt at the surface of the soil?

Try to find some soil where water stands for a long time after a rain. Place some in the lamp chimney. How long does it take for water to percolate down through it?

CRYSTALS

Investigation: How can we grow crystals? (I)

Background Information: Many rocks contain crystals. The shape and color of the crystal are important features that help us to identify minerals. Each kind of crystal will always have distinctive characteristics. With some care and patience, children can grow fairly large and distinctive crystals.

Crystals will be deposited out of saturated solutions as the solution cools and liquid evaporates. In rocks, large crystals indicate very slow cooling, and usually that cooling took place deep underground. Glassy rocks or rocks with very small crystals indicate rapid cooling, and that cooling probably took place at or near the surface. Hence, a student of minerals can tell something about how and where rocks were formed by examining the crystals.

Children can grow crystals using such substances as salt, copper sulfate, or alum.

Materials: A substance such as salt (the coarse kind used to sprinkle on icy surfaces or to make ice cream works better than table salt), copper sulfate, or alum

Two beakers or pyrex containers

Hot plate or other source of heat

One flask or bottle for storing stock solutions

Vaseline

Clear nail polish

Procedure:
1. Dissolve as much of the chemical substance as possible in hot water.
2. Filter the hot liquid to remove any dirt or other foreign substance. (If the chemicals are fairly clean, this step may not be necessary.)
3. Allow the hot, saturated liquid to cool. Usually some crystals will be formed on the bottom.
4. Pour the liquid solution into the flask or bottle. This liquid is the stock solution used to replenish the liquid in the evaporating container.
5. Fill a beaker or other container about half full with the stock solution.
6. Select the best-formed crystal and drop it into the soltuion in the beaker. This crystal is the seed crystal that we hope will grow.
7. Smear a ring of vaseline around the inside of the beaker just above the level of the solution. The vaseline will keep the solution from "creeping" up the sides of the container. Pour in a little more stock solution to raise the level up to the ring of vaseline. Put a card over the top of the container to keep out dust.

8. As the liquid evaporates, the solution becomes oversaturated, and some of the chemical substance should sink to the bottom and be deposited on the seed crystal. A nearly constant temperature should be maintained.

9. The growth of the crystal should be checked daily. With some kind of forceps, the crystal should be turned onto a different face each day. Abnormal growths or "suckers" should be removed from the crystal. If other crystals are precipitated, they should be removed.

10. When the level of the liquid has been noticeably lowered, a little stock solution should be added. To prevent a showering of additional crystals, add drops of water to the evaporating beaker before adding the stock solution. With a little care, children can grow fairly large crystals.

11. When the crystal has achieved the desired size, remove it and allow to dry on a piece of clean paper.

12. Since some substances change over time when they are exposed to air, spread a coat of clear nail polish over the crystal.

Going Further: Grow crystals of other substances and compare them as to shape and other physical attributes. Compare crystals found in rocks with crystals that are grown.

6

Magnetism
and Electricity

INTRODUCTION

Movements due to magnetic and electrical forces are examples of action at a distance. If a magnet is brought near a steel paper clip or nail, the nail or paper clip will be attracted to the magnet, even though the magnet does not make direct contact with either object. A sheet of paper can be placed between the magnet and the object, and the object will still be attracted. Similarly, if a comb is rubbed with a piece of wool or plastic, small pieces of paper and lint will be attracted. There will be movement, even though the magnet or comb does not make direct contact with the other objects.

Magnetism and electricity have many similarities, yet they are different. Magnets will attract only magnetic materials such as iron or steel, but not materials such as paper, lint, plastic, rubber, or hair. On the other hand, an object such as a comb that has been charged by rubbing with plastic or wool will attract only materials such as hair, rubber, plastic, lint, and paper, and it will usually not attract magnetic materials and metals.

There are important relationships between magnetism and electricity. In this chapter there are activities in which children can use electricity to generate magnetic fields and make magnets and compass needles move. There are also activities in which children can use magnets to generate electricity.

It is very important that children have firsthand experiences in working with electricity and magnetism. Reading about them is fine, but experiencing the feel of attraction from a charged comb, seeing a string of paper clips lifted by a magnet, and seeing "my magnet actually work"—these are experiences every child should have.

MAGNETIC POLES

Investigation: Where are the poles of a magnet?   (P, I)

Background Information: Poles of magnets are the parts of magnets where the magnetic effects seem to to be concentrated. Magnets usually have two poles, and these are called the *north pole* and the *south pole*. If a magnet is suspended so that it is free to turn, one pole will tend to become oriented in the direction of the north magnetic pole of the earth, and therefore this pole is called the north pole. The other pole will be oriented toward the south and so is called the south pole.

Because of the way that they have been magnetized or stored, some magnets will be found to have more than two poles. Actually, these may be considered to be more than one magnet.

One of the best ways to locate the poles of a magnet is to use a magnetic *compass.* The needle of a compass is actually a small magnet that is free to turn. The north pole of the compass needle will be repelled by the north pole of another magnet and attracted to the south pole. Similarly, the south pole of a compass needle will be attracted to the north pole and repelled by the south pole of another magnet. Th basic law of magnets is stated thus: *Like poles tend to repel and unlike poles tend to attract.*

Materials: One or more magnets
Magnetic compass
Steel paper clips

Procedure: 1. Spread some paper clips on the table. Set a magnet down upon the paper clips and then lift it. Are more paper clips attracted to some parts of the magnet than others? Usually, more paper clips will be attracted to the poles than to the other parts of the magnet.

2. Move a magnetic compass alongside a magnet (see Figure 6-1). Note

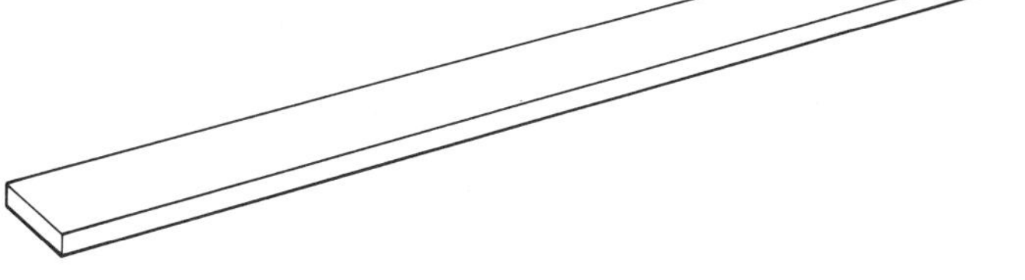

FIGURE 6-1.   Move the compass along the magnet. Are the ends of the compass needle attracted to certain parts of the magnet?

where the compass needle points. You may wish to mark these points on the magnet with chalk or crayon. Where are the places in which the magnetic effects seem to be concentrated? Is it the same end of the compass needle that always points toward these poles?

Going Further: Move the compass alongside magnets of different shapes. How many poles are there in the magnets? Where are they located?

With tape and string, suspend a bar magnet so that it is free to turn. Does the bar magnet line up in the same direction as a compass needle?

FINDING MAGNETS

Investigation: What objects around us are magnets? (P, I)

Background Information: Objects made of magnetic materials such as iron and steel can be magnetized. If they are stored near magnets or remain oriented in the same direction in the earth's magnetic field for some time, these objects will become magnetized.

> The test for whether an object is a magnet is whether one pole of a known magnet, such as a compass needle, will be repelled by some part of the object. Children often believe that the test for a magnet is whether such objects as a paper clip or nail are attracted to it. However, this is not a definitive test because it might be the paper clip or nail that is the magnet.

Materials: Magnetic compass

Procedure:
1. Find a number of objects made of iron or steel, such as scissors, pliers, nails, steel coat hangers.
2. Move the magnetic compass alongside each of the objects. Is one end of the compass needle repelled by some part of the object? If it is, the object is a magnet.
3. Move the compass alongside such objects in the environment as steel chairs or table legs, steel fence posts, and flagpoles. Do any parts of these objects repel one pole of the compass needle?

Going Further: Try to demagnetize an object, such as a nail or paper clip, that has been found to be a magnet. One way is to place the object in an east-west direction and pound it with a hammer. Is the object demagnetized? Another way is to hold it with pliers or tongs in a flame until it gets red hot. After it cools, test it. Has it been demagnetized?

MAKING MAGNETS

Investigation: How can we make a magnet? (P, I)

Background Information: Magnetic materials such as iron and steel can be considered to be made up of many very small magnets called *domains*. When unmagnetized, these small magnets are oriented in all different directions, and their magnetic effects tend to cancel each other. If many of these small magnets can be oriented in a particular direction, then the material may be magnetized. The test for whether an object is a magnet is whether one pole of a known magnet, such as a compass needle, will be repelled by some part of the object.

Materials: Magnet

Magnetic materials such as a nail or bolt

Magnetic compass

Procedure 1. Test the nail or bolt to see if it is a magnet by seeing whether one part of the nail or bolt repels one end of the compass needle. If it does, place the nail or bolt in an east-west orientation and pound it with a hammer until it ceases to be a magnet.

2. Move one pole of a known magnet in one direction alongside the nail or bolt. The stronger the magnet the better. Now test to see whether one end of the compass needle will be repelled by some part of the nail or bolt. Usually, the nail or bolt has to be stroked only once by the known magnet to become a magnet itself. If the nail or bolt is stroked more than once, it should be by the same pole and in the same direction.

Going Further: Fill a test tube with iron filings. Place a cork in the mouth of the test tube and shake. With a magnetic compass, see if any part of the test tube full of iron filings repels one end of the compass needle. Now stroke the test tube with one pole of a magnet (see Figure 6-2). Have the children observe

FIGURE 6-2. With one pole of a magnet, stroke a test tube filled with iron fillings. Does the test tube become a magnet?

what happens to the iron filings. The movement of the iron filings can be compared to those of the domains (small magnets) in an object made of iron or steel. Does any part of the test tube repel one end of the compass needle? If it does, the test tube filled with iron filings has become a magnet.

It is possible to use the earth's magnetic field to make a magnet. Take a bolt or nail that does not repel an end of a compass needle and place it in a north-south orientation. Actually, it is even better if it is tipped a little, because the direction to the earth's magnetic pole is usually at an angle to the surface of the earth. Pound the bolt or nail vigorously so that more of the domains in the object will be oriented in a north-south direction. Move a compass alongside it. Does any part of the bolt or nail repel one end of the compass? If it does, it has become a magnet.

Try to make a magnet having more than two poles. With a pair of pliers, cut a long strip of metal from a wire coat hanger. Stroke one end of the coat-hanger wire in one direction with a pole of a magnet, and the other end with the same pole of the magnet but in the opposite direction. Move a compass alongside the coat-hanger wire. Where are the poles of the coat-hanger wire magnet? What poles are at the ends?

BREAKING MAGNETS

Investigation: What happens when magnets are broken or cut? (P, I)

Background Information: Magnets have at least two poles. Most magnets have a north pole at one end and a south pole at the other end. But, what happens if a magnet is cut or broken? How many poles do the parts have?

Materials: Steel or iron wire coat hanger
Pliers
Magnet
Compass

Procedure: 1. With pliers, cut a long strip of wire from a wire coat hanger.
2. Stroke the wire strip in one direction with one pole of a magnet.
3. Move a compass alongside the wire and note where the poles are. Also note which poles they are. The pole that repels the north end of a compass needle will be the north pole. The pole that repels the south pole of a compass needle be the south pole.
4. With the pliers, cut the coat-hanger wire in half.
5. Move the compass alongside each piece of the wire. Where are the poles? Which poles are they? Does each section have two poles?
6. Cut one of the coat hanger wire sections again and check for the poles.
7. Continue to cut the section of coat-hanger wire and check for the poles. How short can the coat hanger wire be cut and still be able to detect magnetic poles?

Going Further: Place a piece of glass or cardboard on top of a magnet. Sprinkle iron filings onto the glass or cardboard with your finger (see Figure 6-3). Do the iron filings line up in a pattern? Do the patterns indicate where the poles are? Try to keep the iron filings from coming into direct touch with the magnet; once in contact with the magnet, they are difficult to remove.

FIGURE 6-3. Place a piece of cardboard on top of a magnet and sprinkle iron filings onto the cardboard.

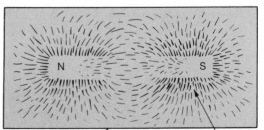

Cardboard ⌐ꜛ   Iron filings ⌐ꜜ

MAGNETISM AND MATERIALS

Investigation: What kinds of materials will magnetism pass through?   (P, I)

Background Information: *Magnetic* materials, such as iron and steel, will be attracted to magnets. *Nonmagnetic* materials, such as paper, wood, and plastic, will not be attracted to magnets.

Magnetism can result in action at a distance. When a paper clip is attracted to a magnet, the magnetic effect takes place through air. In this activity, children will try to find out what materials the magnetic effect will or will not pass through.

> Alnico magnet
> String
> Paper clip
> A variety of materials, such as paper, plastic, leather, wood, iron, steel, copper, or aluminum

Procedure: 1. With a string, hang a magnet from some support, such as a chair or a ruler or stick clamped or weighted down on a table.
2. Tie the end of a string to a paper clip. Tape or tack the string so that the paper clip is suspended in space beneath the magnet, as in Figure 6-4.

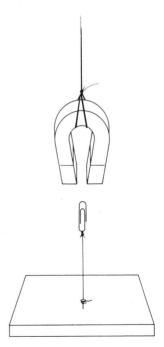

FIGURE 6-4.   Pass a variety of materials between the magnet and the paper clip.

3. Pass a variety of materials (paper, plastic, leather, wood, copper, and aluminum) between the suspended magnet and the paper clip. Is the paper clip affected?
4. Pass objects made of iron or steel, such as scissors and screwdrivers, between the magnet and the paper clip. Is the paper clip affected? Do you feel anything?

Going Further: What kinds of materials seem to have an affect upon the magnetic field around a magnet?

Try to design an experiment to see if water will affect the magnetic field.

A MAGNET MOTOR

Investigation: How can we make a simple magnet motor?   (P, I)

Background Information: The basic law of magnets is that unlike poles tend to attract and like poles tend to repel. This characteristic of magnets is used to build electric motors. Essentially, an electric motor is one magnet turning near another. In most electric motors, the polarity of one of the magnets is changed by reversing the direction of flow of an electric current. When the direction of the flow of current is changed, the polarity of the electromagnet is reversed. If the end of the turning electromagnet had been attracted to the nearby magnetic pole, it will be repelled when the flow of current is reversed. This will cause the electromagnet to begin to spin. In the "magnet motor" in this activity, one of the magnets is moved manually to make the other magnet turn.

Materials: Two bar magnets
String
A support from which to hang one of the magnets

Procedure: 1. Using a string and a support, hang one of the bar magnets so that it is free to turn (see Figure 6-5).
2. Bring one end of the other bar magnet near one end of the suspended magnet. Coordinate the movement of the bar magnet in your hand so that the suspended magnet continues to turn, making a magnet motor.

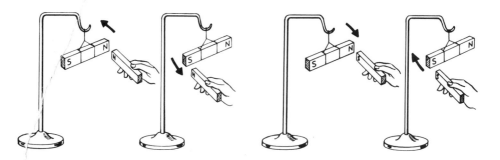

FIGURE 6-5.   A magnet motor in which like poles repel and unlike poles attract.

Going Further: Using a bar magnet, make a compass needle turn on its axis.
Open up an electric bell and try to find out how it works. Since it uses electricity to generate motion, it is an electric motor.

AN ELECTROMAGNET

Investigation: How can we make an electromagnet? How can we change the poles of an electromagnet?   (P, I)

Background Information: When an electric current flows through a conductor, a magnetic field is generated around the conductor. The strength of this magnetic field can be increased by winding the wire into a coil around such magnetic material as an iron or steel nail or bolt.

Materials: Insulated wire
Iron or steel nail or bolt
Dry cell
Switch
Paper clips
Compass

Procedure: 1. Wrap the insulated wire around the nail or bolt.
2. Attach the ends of the wire to the poles of a dry cell (see Figure 6-6). While not absolutely necessary, it is desirable to connect a switch into the circuit so that the circuit can quickly be broken. Children should be cautioned not to leave the switch closed for very long because this causes a great drain on the dry cell.

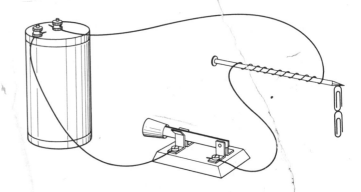

FIGURE 6-6.   An electromagnet can be made by winding insulated wire around a nail or bolt.

3. Dip one end of the electromagnet into a pile of paper clip. Are the paper clips attracted to the end of the electromagnet?
4. Hold a magnetic compass near one end of the electromagnet and note which pole of the compass needle is attracted.

us

5. Change the connections of the ends of wire to the dry cell. This will reverse the direction that the electric current flows through the electromagnet.

6. Hold the magnetic compass near the same end of the electromagnet. Now which pole of the compass needle is attracted?

Going Further: Study the relationship between the number of turns in the wire coil and the strength of the electromagnet. See how many paper clips can be held in a chain at the end of the electromagnet. Now increase the number of turns in the wire coil. Does this affect the number of paper clips that can be held in a chain?

MAKING A PAPER ELECTROSCOPE

Investigation: How can we detect electric charges? (P, I)

Background Information: The *electroscope* is a device used to detect electric charges. Very fine electroscopes are made with thin gold leaf; others are made with aluminum leaf. However, in order to detect electric charges it is possible to use a paper electroscope.

The electroscope operates on the principle that objects that have like electric charges will tend to be repelled, and those with unlike charges attracted.

Materials: Strip of paper (The paper strip can be torn from sheets of newspaper. Newspaper that is grayish in color seems to work best. This grayish newspaper contains less filler.)

Wooden rod (A broom handle, dowel rod, or meter stick will work.)

Procedure: 1. Tear a strip of paper about 5 centimeters (2 inches) wide from a newspaper. The strip can be torn across the width of the spread-out paper and across the centerfold.
2. Hang the strip of paper from the wooden support. Often, it is helpful to have the strip of paper hanging so that the center fold is over the wooden rod.

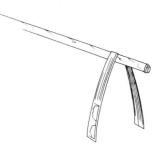

FIGURE 6-7. A paper electroscope can be made by hanging a strip of newspaper over a wooden support.

3. With one hand, hold the paper strip so that it will not pull off the wooden rod. With the other hand, place your index finger between the strips of paper with your thumb on one side and the middle finger on the other to charge the paper. Pull the fingers down while grasping the sheets of paper lightly. Usually, this will charge the sheets of paper, and they will fly apart.
4. Rub a comb with a piece of wool and put the comb between the strips of paper. What happens to the strips of paper?
5. Rub the comb with a sheet of plastic and place the comb between the strips of paper. Now what happens to the strips of paper?

6. Rub the comb with a variety of materials. With what materials will the comb cause the newspaper strips to diverge farther? With what materials will the strips come together?

7. Rub a glass object with various materials and place it between the charged newspaper strips. When do the strips diverge? When do they come together?

Going Further: Obtain an electroscope with aluminum or gold leaves and try some of the same experiments with it.

DETECTING ELECTRIC CURRENTS

Investigation: How can we make a galvanometer?   (P, I)

Background Information: A *galvanometer* is a device for detecting an electric current. It works on the principle that an electric current generates a magnetic field. This magnetic field will affect a nearby magnet.

The galvanometers that will be made consist of a coil of wire through which the electric current will flow and a sensitive compass needle that will move when it interacts with another magnetic field.

Materials: Insulated wire
Compass
Dry cell

Procedure: 1. Wind insulated wire into a coil around a compass. This makes a simple galvanometer.
2. Touch the ends of the wire to the terminals of the dry cell (see Figure 6-8). What effect does this have on the compass needle?

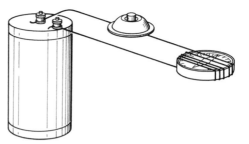

FIGURE 6-8.   A galvanometer for detecting electric currents can be made by winding insulated wire around a compass.

Going Further: An even more sensitive galvanometer can be made by floating a magnetized needle in water. Wind a coil of wire around a small glass dish (a petri or other glass dish can be used). Cover the bottom of the dish with water. Magnetize a steel needle by stroking it in one direction with a pole of a magnet. Float the needle on top of the water. Gently drop the needle flat onto the water or lower it onto the surface of the water with a cradle made of thin wire (see Figure 6-9). The needle will float on the surface film caused by the surface tension of the water. If difficulty is encountered, the needle can be pushed through a small piece of cork and floated. Touch the ends of the coil of wire to the terminals of the dry cell. What happens to the needle?

The glass dish can be placed on an overhead projector, and an entire class can see the movements of the needle projected onto the screen. These galvanometers can be used in other science activities to detect electric currents.

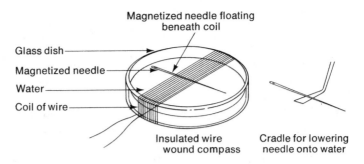

Glass dish

Magnetized needle

Water

Coil of wire

Magnetized needle floating
beneath coil

Insulated wire
wound compass

Cradle for lowering
needle onto water

FIGURE 6-9.   A sensitive galvanometer can be made by floating a magnetized needle below a coil of insulated wire.

ELECTRICITY FROM SALT AND LEMONS

Investigation: How can we use salt or lemons or other chemicals to generate electricity? (P, I)

Background Information: Three materials are needed to generate electricity with chemicals: two dissimilar electrodes and an electrolyte that will react with the electrodes. The *electrodes* are usually two different metals, such as zinc and copper. The *electrolyte* is a substance that will react with the electrodes; often it is a salt solution or an acid, such as citric acid.

    The electric cell generates an electric current when the electrolyte reacts faster with one electrode than with the other. Citric acid or salt will react faster with zinc than with copper. The electric current that is generated can be detected with a galvanometer.

Materials:
Strip of copper or copper coin
Strip of zinc (Zinc can be obtained from the outer metal case of a dry cell.)
Knife
Lemon
Salt
Two different coins
Paper toweling
Tumbler
Copper wire
Galvanometer

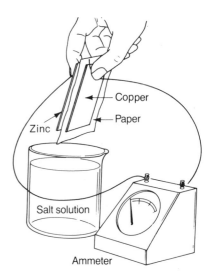

FIGURE 6-10. Electricity is generated when a strip of paper dipped in a salt solution is placed between copper and zinc.

Procedure: 1. Pour some salt into water in a tumbler and stir.

2. Dip a piece of paper toweling into the salt solution and place the paper between a strip of copper and a strip of zinc.

3. Connect one terminal of a galvanometer to the copper strip, and touch a wire from the other terminal to the zinc (see Figure 6-10). Is there an indication that an electric current has been generated? (To prevent damage if a commercially made galvanometer is used, the wire from one terminal should be just touched to the zinc strip and removed quickly if the needle swings sharply to the edge of the dial.)

Going Further: Place the paper dipped in the salt solution between two different coins. Connect the coins to the galvanometer. Is an electric current generated?

With a knife, cut two slits in a lemon. Stick a strip of copper in one slit and a strip of zinc in the other. Connect the strips to a galvanometer. Is an electric current generated?

Place an old dry cell in a vise and saw it open with a hacksaw. Try to find the two electrodes (zinc can and carbon rod) and the electrolyte (the black material).

FLASHLIGHT

Investigation: How does a flashlight work? What is wrong with the flashlight?   (I)

Background Information: A flashlight essentially consists of a bulb, one or more dry cells, a conductor to connect the base of a dry cell to the outside terminal of the bulb, and a switch. In this activity, children make such a flashlight.

For the flashlight to light, there must be a complete circuit. In most flashlights, the center post of one dry cell touches the center terminal of the bulb. Often there are one or more additional dry cells whose center post touches the base of the dry cell ahead of it. At the base of the flashlight there is a spring that keeps the dry cells pressed together and the center post of the foremost dry cell pressed against the center terminal of the bulb. The spring is also part of the electrical circuit; it makes contact with the base of the last dry cell and connects with the conductor that leads to the switch. Usually the switch makes and breaks a contact with the outer terminal of the bulb.

After the children understand how a flashlight works, they can be given a series of problems to see if they can diagnose the malfunction, correct it, and make the flashlight work.

Materials: Flashlight
Two or more dry cells that fit the flashlight
Flashlight bulb
Insulated copper wire

Procedure: 1. Make a simple flashlight (Figure 6-11). Wrap the bared end of a short piece of copper wire around the outside terminal of a flashlight bulb.

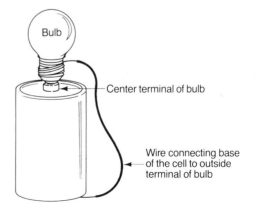

Bulb

Center terminal of bulb

Wire connecting base of the cell to outside terminal of bulb

FIGURE 6-11.   A simple flashlight can be made by pressing the base of the bulb against the center post of the dry cell and connecting the outside terminal of the bulb to the bottom of the dry cell.

Press the center terminal of the flashlight bulb against the center post of a dry cell. Press the base of the dry cell against the other end of the short piece of wire. This is a simple flashlight, and if the dry cells and bulb are in good order, the bulb should give off light.

Have the children note the parts of this simple flashlight. What happens if the circuit is broken? How can the flashlight be made to light again?

2. Examine a flashlight. Find the essential parts. Give the children a flashlight. Have them take it apart. Ask them to find answers to such questions as:

How is contact made with the center terminal of the bulb?

How do the dry cells make contact with each other?

How is contact made with the base of the last dry cell?

How is the connection made between the base of the last dry cell and the outside terminal of the bulb?

How is the circuit in the flashlight made and broken?

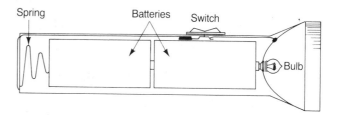

FIGURE 6-12. Have the children trace the electric circuit in the diagram and find the parts of the circuit in a flashlight.

Going Further: Children can use their knowledge of how a flashlight works to diagnose why a particular flashlight does not work, and then proceed to repair it. Many homes and schools have flashlights that are not working that children can try to repair.

"What is wrong with the flashlight?" Here are some puzzles. Make the following adjustments in a flashlight, and then have the children try to find out why it does not work and fix it so that it does work:

1. Turn one of the dry cells around so that the bases of two dry cells are touching.
2. Put a piece of tape over the center post of one of the dry cells.
3. Remove or place a piece of tape over the end of the spring in the base.
4. Bend the brass conductor from the switch so that it does not make contact with the outside terminal of the flashlight bulb.
5. Replace the good bulb with a bad one.
6. Replace a good dry cell with a bad one. Are the children able to find out what is wrong with the flashlight and repair it?

USING A MAGNET TO GENERATE ELECTRICITY

Investigation: How can we use a magnet to generate electricity? (I)

Background Information: When an electrical conductor, such as a copper wire, moves through a magnetic field, an electric current is generated in the conductor. This is the scientific principle that is used in the huge generators that generate most of our electricity.

Materials: Insulated copper wire
Magnet
Galvanometer

Procedure: 1. Wind the copper wire into a coil and attach the ends of the wire to a galvanometer (see Figure 6-13).

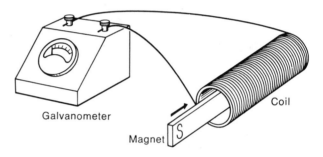

FIGURE 6-13. An electric current is generated when a magnet and a coil of wire move relative to each other.

2. Move the magnet in and out of the coil of wire. Is there an electric current generated in the coil of wire?

Going Further: Hold the magnet in the coil of wire and move the two together. Is there a current generated? This can be used to develop a concept of relative motion. The magnet and the coil must move relative to each other for an electric current to be generated.

ELECTRIC MOTOR

Investigation: How can we build an electric motor? (I)

Background Information: An electric motor is a device for converting electrical energy to mechanical energy. As such, it is the opposite of an electrical generator, which converts mechanical energy to electrical energy.

An electric motor is essentially one magnet revolving near other magnets. At least one of these magnets must be an electromagnet. In the motor that is described, the magnet that rotates is an electromagnet. This rotating electromagnet is often called the *armature*. Brushes make sliding contact with the ends of the armature wire. In this motor, the field magnets are permanent magnets.

The electric motor operates on the principle that like magnetic poles repel and unlike magnetic poles attract. The polarity of an electromagnet can be reversed by reversing the direction of the flow of electricity. The polarity of the electromagnet (armature) reverses as the armature rotates and the ends of the armature wire make successive contact with each of the two brushes.

Materials: Two permanent alnico magnets

Two dry cells (No. 6 dry cells work very well.)

One test tube

Wire coat hanger

Two supports for permanent magnets (oatmeal boxes, empty tennis ball cans, or milk cartons)

Insulated copper wire (Various thicknesses of wire can be used—No. 20 works very well.)

Tape

Pliers or tin shears

File (not absolutely necessary)

Small cork

Procedure: 1. With pliers or tin shears, cut three pieces of the coat hanger, each about 8 centimeters (3 inches) long.

2. Measure and cut off about 4 meters (13 feet) of insulated copper wire for the armature. With the pliers, remove about 8 centimeters (3 inches) of insulation from each end of this wire.

3. With about 10 centimeters (4 inches) of the wire left free at the end, hold the three pieces of coat-hanger wire together and start winding the insulated wire around them. Be sure to *always wind in the same direction*. Wind from the middle of the coat-hanger wires out to one end, back to the other end, and back to the center. Try to distribute

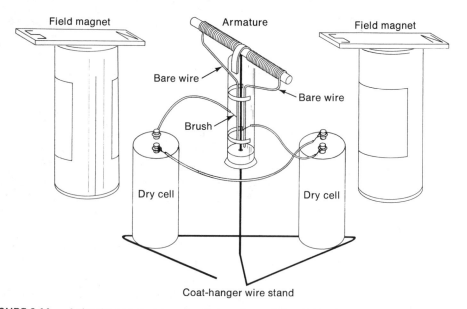

FIGURE 6-14. A sketch of an electric motor that can be made by children.

the turns so that the armature will be well balanced. Leave about 10 centimeters (4 inches) of wire free at the remaining end.

4. With a piece of tape, fasten the middle of the coil to the rounded bottom of the test tube. This coil will rotate and serve as the armature of the motor.

5. Stretch the bared ends of the armature coil along opposite sides of the test tube. Fasten the bared ends in place with a thin strip of tape near the coil and another near the opening of the test tube. The brushes will make sliding contact with the bared ends of the wire.

6. With pliers or tin shears, cut off the hook part of the wire coat hanger. Bend the remainder of the wire coat hanger, as shown in Figure 6-14, so that a length of the wire coat hanger slightly longer than the test tube will be vertical. The test tube will turn on the point of this vertical wire. To reduce friction, file the point of this wire so that it is sharp. Some children have stapled the coat-hanger wire to a broad to make the motor steadier.

7. Insert a cork with a small hole in the center into the opening of the test tube.

8. Put the test tube onto the vertical wire so that the inside of the bottom of the test tube rests on the sharp tip of the vertical wire. Spin the test tube to make certain that it spins freely.

9. Support two alnico bar magnets on tennis ball cans or milk cartons so that they are at the same height as the rotating coil. The magnets

should be opposite each other; the north pole of one magnet and the south pole of the other should be near the coil.

10. Cut two pieces of insulated wire about 20 centimeters (8 inches) long. Remove about 6 centimeters (2½ inches) of insulation from one end and 2 centimeters (1 inch) from the other end of each wire. These two wires will serve as the brushes.

11. Connect one bared end of another piece of wire to the center post of one dry cell and the other bared end to the outside post of a second dry cell.

12. Connect the short bared end of one of the wires that will serve as a brush to the outside post of one dry cell, and the short bared end of the second wire brush to the center post of the second dry cell.

13. Arrange the dry cells so that the long bared sections of the brushes will brush against the test tube as it rotates. A small piece of tape can be wound around the wires so that they will be held the right distance apart to continue to make contact with the test tube and the ends of the armature wire.

14. The brushes should be positioned so that they will make contact at the same time with the bared wires on the sides of the test tube; if positioned correctly, an electric current will periodically flow through the armature, making it an electromagnet. The brushes should make contact with the wires when the ends of the electro-magnet (armature) are about one-third of the way past the nearest pole of the field magnets. The exact points of contact between the brushes and the wires on the test tube usually have to be adjusted to get the armature to rotate as fast as possible. This can be done by moving the dry cells.

15. Give the armature a spin. Do you see any sparking as the brushes make contact with the bared ends of the armature wire? Experiment with adjusting the positions of the brushes or the field magnets to make the armature turn faster.

Going Further: Move the magnets into the position in which the motor is running fastest. Remove one of the magnets. Does the motor still run? Change one of the magnets so that both of the field magnets have the same pole nearest the armature. Now does the armature rotate?

Note the direction the armature is spinning. Change the magnets so that the north pole is where the south pole was, and the south pole where the north pole was. Now which way does the armature spin?

Again, note the direction the armature is spinning. Move the dry cells so that the brush from the outside post is where the brush from the center post was and the brush from the center post is where the brush from the outside post was. Now which direction does the armature turn? Suggest and try ways to make the armature spin faster.

7

Machines

INTRODUCTION

In this chapter we deal with some of the most basic laws in science. Too often, these laws are stated in textbooks, discussed, and perhaps memorized, but they do not take on real meaning in terms of the phenomena of the world in which we live. Children can develop a better and deeper understanding by actually taking part in science activities that give them firsthand experiences with the application and meaning of these laws.

Sometimes science activities are carried out without developing an understanding of very basic scientific laws and principles. While there is nothing wrong with having fun with science, it is still possible to learn while having fun. Children should be helped to generalize from science activities. For example, balancing weights can be fun, and in some situations, perhaps that is sufficient. But this activity provides an opportunity to learn the basic principle of simple machines, and this principle can be used in the study, use, and construction of many different kinds of machines.

The following are among the basic scientific laws and principles developed and applied in this chapter:

1. Measurement is comparison. The simplest form of measurement is direct comparison.
2. Standard units of measurement are arbitrary, and others can be used.
3. The masses of objects can be compared using an equal-arms balance. An equal-arms balance can be used as a physical model of an algebraic equation, and to introduce children to equations.
4. Objects at rest tend to stay at rest unless acted upon by some force.
5. The force times the distance the force moves equals the resistance times the distance the resistance moves.
6. Machines can help us to change the direction a force is applied, increase the amount of force that can be applied, reduce the force that is applied, and apply a force in ways that cannot be done without machines.
7. The mechanical advantage of a machine is the force exerted by the machine over the force applied to the machine.
8. The theoretical mechanical advantage of a machine such as the inclined plane is distance the force is applied over the height the weight is lifted.
9. The force of friction depends upon the nature of the surfaces in contact and the amount of force pressing them together.

MEASUREMENT

Investigation: How can we use unconventional units such as paper clips for measurement? (P, I)

Background Information: Essentially, measurement is comparison, and the simplest form of measurement is direct comparison. For example, two children can stand back-to-back to see directly which child is taller. The acceptance of units of measurement such as centimeters and meters allows us to make comparisons when the objects cannot be compared directly. If we announce that Cecilia is 150 centimeters tall, this statement can be understood by children in Argentina, Britain, Chile, Denmark or anywhere; the centimeter is a *standard unit* of measurement that is understood and can be used for measurement around the world.

However, the *standard units* of measurement that we use are arbitrary, and others could be accepted and used. Since it is important that children recognize that standard units are arbitrary and that others can be used, we will use paper clips as units of measurement in this activity.

Materials: Paper clips

Procedure:
1. Hook 10 paper clips together. This unit of length can be called a "decaclip." By using a decaclip we can quickly find the length of an object that is 10 paper clips long.
2. Using decaclips and single paper clips, measure the length of a table in paper clips.
3. Measure the width of the table in paper clips.
4. By multiplying length times width, find the area of the table top in square paper clips.
5. Use the decaclips and single clips to find the length, width, and area of other objects in the room.

Going Further: Use common objects or containers as standard units in the measurement of weight and volume. For example, steel washers and nails can be used as units of weight, and juice cans can be used as units of volume.

BALANCE

Investigation: How can we make an equal-arms balance?   (P, I)

Background Information: There are many situations in which we have to compare or measure the weight or mass of objects. We can compare the weights of two objects by "hefting" them in our hands. If the difference in the weights of the two objects is considerable, we can easily tell which object is heavier. But when the differences are small, we cannot compare effectively by hefting.

The masses or weights of objects can be compared with an *equal-arms balance*, such as the one shown in Figure 7-1. Such a balance can also be used to compare the mass of an object with standard units of mass, such as grams and kilograms (or paper clips or nails).

With older children, the equal-arms balance can be used as a physical model of an algebraic equation. We can physically show that for an equation to balance, operations on one side have to be balanced by similar or equal operations on the other side.

Materials: Half-meter stick or other stiff rod
Three spring clamps
Two small pans or cups
String
Paper clip
Support for the equal-arms balance

Procedure: 1. Assemble the equal-arms balance as shown in Figure 7-1. In the illustration, the rod or half-meter stick is shown suspended from a nail in

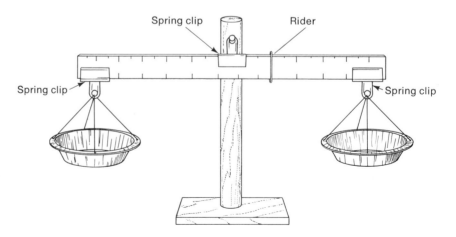

FIGURE 7-1.   An equal-arms balance can be made with a half-meter stick or stiff rod, three spring clips, two aluminum pans, and a wire rider.

an upright stand. If this is difficult to construct, it can also be hung between two tables or chairs.

2. A rider can be made by unbending a paper clip and hanging it on the equal-arms balance. Balance the rod by sliding the rider along the rod until the rod hangs horizontally.

3. Take two objects that seem to be of equal weight when hefted and place one into each pan of the equal-arms balance. Which is heavier?

4. Place an object, such as a small ball, in one pan. Add paper clips to the other pan until the pans are in balance. What is the weight of the object in "paper clips"?

Going Further: Use the equal-arms balance as a physical model of an equation. Raise such questions as the following to the children:

1. The pans are in balance. We add 8 paper clips to one pan. How many paper clips will have to be placed in the other pan to achieve a balance?

2. If we have 13 paper clips on one side and 8 paper clips on the other side, how many paper clips will have to be added to the other side to make them balance?

3. If we have 15 paper clips on one side and 9 paper clips on the other, how many paper clips will have to be removed from one side to make the sides balance?

CENTER OF GRAVITY

Investigation: Will it tip? (P, I)

Background Information: The center of gravity of an object is the point at which the object could be suspended so that it would be in balance.

The location of the center of gravity will determine whether an object will tip or remain at rest. If an imaginary vertical line down from the center of gravity of an object passes through the base of the object, it will be stable and will not fall. For example, if a coin is placed on edge, it will remain on edge as long as a line from its center of gravity goes through the base that touches the supporting surface. But if the coin is tipped just a little bit, it will fall over.

Materials:

Coin	Ruler	Pin
Cork	Wire	String
Two Forks	Piece of cardboard	Weight
Hammer	Nail	Pencil

Procedure:
1. Place a coin on edge. What happens if the coin is tipped a little?
2. Lay a coin flat on the table. Does this coin tip as easily?
3. Stick the tines of two table forks into a cork as shown in Figure 7-2.

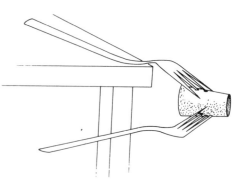

FIGURE 7-2. What keeps the forks from tipping and falling?

Place the handle of one fork on the top of a table so that the second fork hangs below the table. Do the cork and forks tip and fall to the floor? Why?

4. Make a wire loop and use it to attach a hammer to a ruler as in Figure 7-3.

Place the ruler near the edge of a table top so that the head of the hammer is beneath the table top. Do the ruler, hammer, and wire tip and fall to the floor? Why?

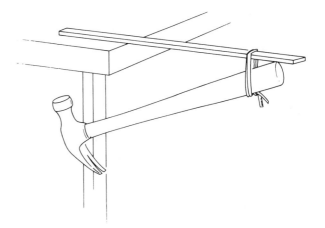

FIGURE 7-3. Why don't the hammer and ruler tip and fall?

Going Further: Cut an irregular piece of cardboard. With a nail, make holes as shown by A, B, C, D in Figure 7-4.

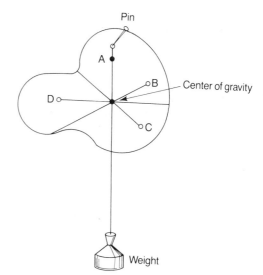

FIGURE 7-4. Children can find the center of gravity of a piece of cardboard.

Place a pin through one of the holes and into a cork or bulletin board. With a string, suspend a weight from the pin. Draw a pencil line on the cardboard using the string as a guide. Repeat with the pin in each of the holes after allowing the piece of cardboard to swing to rest. The point where the lines cross should be the center of gravity of the piece of cardboard. If a pin is pushed through the cardboard at that point, does the cardboard tend to turn? Why?

BALANCE OF FORCES

Investigation: How far can we pull one side of a towel down across a towel rack before the whole towel slides off?   (P, I)

Background Information: Objects at rest tend to stay at rest unless acted upon by some force. Usually a towel will hang on a rack without moving. However, if the force on one side of a towel becomes greater than the force on the other side plus the force of friction, we expect the towel to move because of the imbalance of forces.

Among other factors, the force of friction depends upon the nature of the surfaces in contact. Therefore, we would expect that the distances a towel can be pulled before it slides will vary with the nature of the towel and the nature of the surface of the towel rack.

Perhaps the most important concept to be developed through this activity is: If there is a balance of forces, there is no motion; an imbalance of forces can lead to motion.

Materials: Towels and other materials
Towel rack or bar across which towels can be hung
Measuring stick or tape

FIGURE 7-5.   How far can the towel be pulled before it begins to slide?

Procedure: 1. Measure the length of a bath towel.
2. Hang the towel across a bar, stick, or bath rack. A broom handle works beautifully.
3. Pull one end of the towel until it begins to slide. With a measuring stick or tape, measure the length of the short end of the towel just as the towel begins to slide.
4. Calculate the ratio of the length of the short section of the towel when the towel starts to slide to the length of the long section.

$$\frac{\text{length of short section of towel}}{\text{length of long section of towel}} =$$

5. Try other kinds of toweling material and cloth. Does the nature of the material affect the distance the material must be pulled before it slides?

6. Try different surfaces, such as metal and plastic, for towel racks. Does the nature of the towel rack surface affect the distance the material must be pulled before it slides?

Going Further: Try pulling the towel down at an angle. How do you account for what happens? Try to find other situations in which an imbalance of forces leads to motion.

BALANCING WEIGHTS

Investigation: Where should the weight be placed to balance another weight? (I)

Background Information: The basic principle of simple machines is that *the force times the distance the force moves equals the resistance times the distance the resistance moves.* With levers, a slight variation of this can be used:

force X force arm = resistance X resistance arm.

If the force and the resistance are equal and at equal distances from the *fulcrum* (the pivot point), they will balance (see Figure 7-6). In practice a number of small errors can enter into the experiments to test this principle.

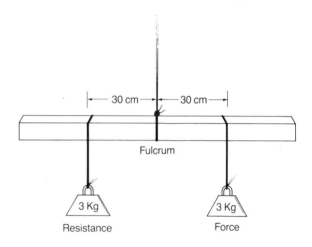

FIGURE 7-6. If the force and resistance are equal and at equal distances from the fulcrum, they should balance.

If the force arm is twice as long as the resistance arm, then a force only half as great as the resistance should balance the resistance (Figure 7-7).

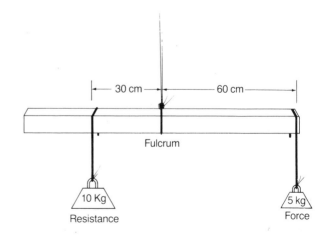

FIGURE 7-7. A force can balance a weight twice as large if the force arm is twice as long as the resistance arm.

A very sensitive device for carrying out experiments to test the basic principle of machines can be made with a ruler and paper clips. This activity gives children practical experience with the concept of an equation.

Materials: Ruler (A half-meter for one-foot ruler works well.)

Paper clips

Support from which ruler can be suspended (A long pencil or dowel stick works well.)

Procedure: 1. With a string or paper clip fastened near the center of the ruler, suspend the ruler from the support. The string can be tied around the ruler, however, many rulers have a hole at the center or one can be drilled. The hole is helpful in suspension.

2. Use a paper clip as a rider. Squeeze the paper clip so that there is some friction between the paper clip and the ruler, and so that the rider will not easily slide on the ruler. Move the rider along the ruler until the ruler is balanced.

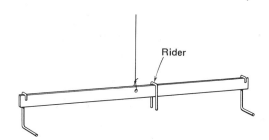

FIGURE 7-8.   A device for balancing weights.

3. Hang 4 paper clips from a hook at a point 5 centimeters or 2 inches from the fulcrum. Where should 2 paper clips be hung to balance the 4 paper clips?

4. Slide the clip holding the two paper clips along the ruler until the ruler is in balance. Is it near the point predicted by use of the basic principle of simple machines?

$$4 \text{ clips} \times 5 \text{ cm} = 2 \text{ clips} \times X \text{ cm}$$

5. Try placing various combinations of clips at various places on the ruler and having the children predict where other combinations of clips should be placed to have the ruler in balance.

Going Further: Another approach is to have children predict how large a weight has to be placed at a given point to balance a given weight at another point.

SIMPLE MACHINES

Investigation: What are some simple machines? What are some of the advantages of using simple machines? (P, I)

Background Information: There are a variety of simple machines that we use in our everyday lives, such as levers, pulleys, inclined planes, and wheels and axles.

Among the advantages of using simple machines are: changing the direction a force is applied; increasing the amount of force that can be applied; reducing the force that is applied; applying the force to a place that cannot otherwise be reached; and applying a force in ways that cannot be done without machines.

Materials: A room such as a classroom or living room

Samples of simple machines are helpful but not absolutely necessary

Procedure: 1. Have the children try to find as many examples of simple machines as possible. The following are some machines that can be found in a room:

Pulley system in Venetian blinds

Wheel and axle in Venetian blinds

Stairs and ramps as inclined planes

Electric switches

Door knobs

Almost any tools and kitchen apparatus

2. After the children have found several simple machines, have them identify the kind of machine that it is. Remember, some may be combinations of more than one kind of simple machine.

Going Further: Try finding simple machines in other kinds of rooms, in an automobile, on a farm, and so on.

LEVERS

Investigation: How can we lift objects with levers? (P, I)

Background Information: A lever consists of a stiff bar of some kind and a *fulcrum*, or point of support on which it pivots. However, there are three different classes of levers, depending on the relative positions of the weight and the force applied. These different classes of levers are shown in Figure 7-9.

Class 1
Fulcrum in middle

Class 2
Weight in middle

Class 3
Force in middle

FIGURE 7-9. Three classes of levers.

Materials: Ruler or stiff bar
Small weight

Procedure: 1. Have the children examine Figure 7-9. Then ask them to use the ruler and the small weight to demonstrate each of the classes of levers.

2. Examine some common utensils and tell which class of lever they are. The following are a few examples of each class:

Class 1: Scissors and pliers—each consists of two Class 1 levers.

Class 2: Many nut crackers consist of two Class 2 levers.

Class 3: Some tongs and a shovel might be considered Class 3 levers. A pencil also can be a Class 3 lever.

Going Further: Try to find examples of machines in the room. Try to categorize each of the machines as to which class it belongs to.

PULLEYS AND MECHANICAL ADVANTAGE

Investigation: How can we use pulleys to move objects?   (I)

Background Information: The *mechanical advantage* of a machine is a comparison of the force exerted by the machine and the force applied.

$$\text{Mechanical Advantage (M.A.)} = \frac{\text{Force exerted by the machine}}{\text{Force applied to the machine}}$$

For example, if a machine can make it possible to exert a force of 6 pounds when a force of 2 pounds is applied, the mechanical advantage of the machine will be 3.

$$\text{M.A.} = \frac{6 \text{ lbs}}{2 \text{ lbs}} = 3$$

In pulley systems, the mechanical advantage is roughly equal to the number of supporting strands.

However, a price is paid for mechanical advantage in that the greater the mechanical advantage, the greater the distance the force has to be applied in order to move a resistance a certain distance. For example, if the mechanical advantage is 3, the force has to be exerted through 3 times the distance the resistance is moved.

Materials: Measuring stick
Pulleys
String
Weights
Spring scale

Procedure: 1. Arrange the string, pulley, and weight as shown in Figure 7-10, where there is one supporting strand.
2. With a spring scale, measure the force that is needed to support the weight. The force applied is measured in terms of the distance the spring is extended. How does this force compare with the weight of the object? What is the mechanical advantage of this pulley system?
3. With a ruler, measure the distance a force has to be applied to lift a weight a given distance. This can be done by measuring the distance the end of the string was pulled. For example, how far must a force be applied to lift an object 10 centimeters?

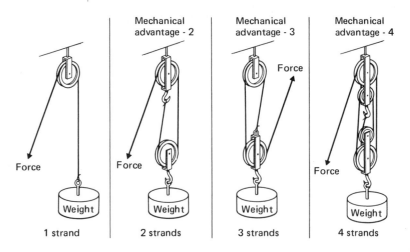

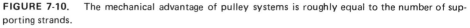

FIGURE 7-10. The mechanical advantage of pulley systems is roughly equal to the number of supporting strands.

4. Arrange pulley systems that have 2 supporting strands, 3 supporting strands, and 4 supporting strands (see Figure 7-10). For each system:

Measure the force that is needed to support the weight.

Calculate the mechanical advantage.

Measure the distance a force must be applied to lift an object a given distance.

Going Further: Examine each of the pulley systems and count the number of supporting strands. The theoretical mechanical advantage is roughly equal to the number of supporting strands. In each case, compare the actual mechanical advantage that has been found with the theoretical mechanical advantage. What might lead to differences between the two?

INCLINED PLANES AND EXERTING A SMALLER FORCE
OVER A GREATER DISTANCE

Investigation: How can we use a relatively small force to lift a greater weight? (I)

Background Information: An inclined plane is a ramp or other incline that can be used to help lift a heavy object. A ramp makes it possible to drive a car or truck up a certain height. A stairway is also an example of an inclined plane.

With an inclined plane, it is possible to spread the application of a force over a longer distance. The *theoretical mechanical advantage* of an inclined plane is the length of the plane over the height the object is lifted.

$$\text{T.M.A.} = \frac{\text{Length of plane}}{\text{Height}}$$

The *actual mechanical advantage* is the weight of the object lifted over the force exerted.

$$\text{A.M.A.} = \frac{\text{Weight}}{\text{Force exerted}}$$

Materials: A board or plank several feet long

A support for one end of the board (Books, large blocks, or a chair will do.)

A toy cart or truck that rolls easily

Spring scale

String

Measuring tape or stick

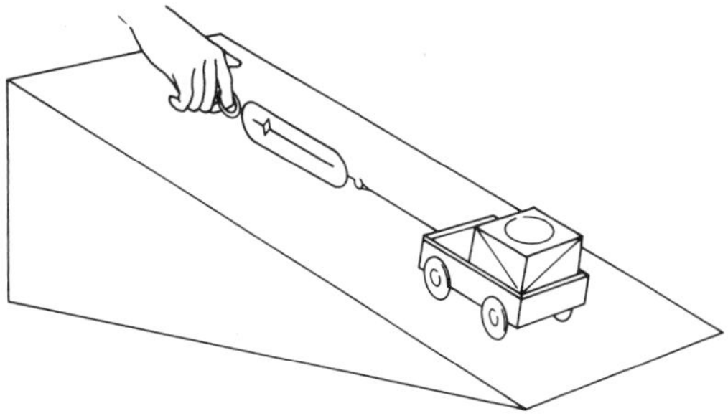

FIGURE 7-11. How much force is needed to pull a cart and a weight up an inclined plane at a constant velocity?

Procedure: 1. Using a board and a support, set up an inclined plane somewhat like the one shown in Figure 7-11.

2. Measure the length of the board and the height the weight is to be lifted. Divide the length of the board by the height to be lifted. This will give the theoretical mechanical advantage.

3. With the spring scale, measure the weight of the small cart or truck and the object to be lifted. Add the two weights to give the total weight to be lifted.

4. Make a loop in one end of a short string and attach the other end to the cart. Hook one end of the spring scale through the loop.

5. Using the spring scale, pull the cart at a constant velocity up the inclined plane. (It is very important that the cart be pulled at a steady, constant velocity.) How much force is needed to pull the cart and the object up the inclined plane?

6. Divide the combined weight of the cart and the object by the force that has to be exerted to move them up the inclined plane to give the actual mechanical advantage. Compare with the theoretical mechanical advantage.

Going Further: Try to determine the theoretical mechanical advantage of a staircase. The height can be determined by measuring the height of a single step and multiplying by the number of steps. The length can be measured by stretching a measuring tape up the stairs from the bottom to the top, or by measuring the width of a stair step and multiplying by the number of steps. How much easier is it, theoretically, to walk up the stairway than to climb the same height without this inclined plane?

Try to find examples of inclined planes. Note that a screw is really an inclined plane curving around an axis in the same way that a spiral stairway is a curved inclined plane.

FRICTION

Investigation: How does the force of friction change?   (I)

Background Information: There is friction in all machines. In many cases, we try to reduce the amount of friction by using rollers or ball bearings. Oil and other lubricants are also used to reduce friction.

Friction is also necessary. Walking, for example, would not be possible without friction. Friction is used in brakes to stop bicycles and automobiles.

In general, the force of friction increases as the force pressing two surfaces together is increased. The size of the area of contact usually does not affect the force of friction. The nature of the surfaces in contact does affect the force of friction.

Materials: Board
Small block of wood to which a string can be attached
Weights
String
Spring scale
Sandpaper
Dowels or round pencils

Procedure: 1. Attach a string to a block of wood and hook a spring scale through the loop in the other end of the string.

2. Pull the block across the board several times. In carrying out these experiments, the block of wood should be pulled across the board at a constant velocity. With the spring scale, measure the force needed to pull the block each time (see Figure 7-12). Ask the children which of the several measures derived should be used for the force necessary to pull the block across the board. (They may suggest that the figures be averaged.)

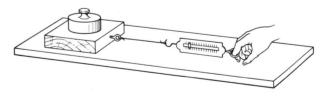

FIGURE 7-12.   Measure the force of friction as you pull the block at a constant velocity.

3. Turn the block on its side and pull it along the board. Compare the forces needed to pull the block across the board when on its side and when flat.

4. Again lay the block flat on the board. Place a weight on the board, and with the spring scale, measure the force necessary to pull the block at a steady, constant velocity across the board. Repeat as additional weights are placed on the board. Record the results on a graph such as the one in Figure 7-13. What relationship does there seem to be between the force of friction and the weight pressing the surfaces together?

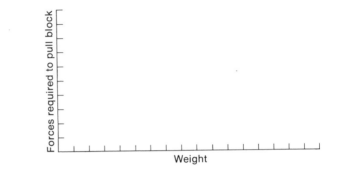

FIGURE 7-13. A graph on which children can enter their findings on the relationship between weights pressing surfaces together and the force needed to pull the block at a constant velocity.

5. Tack a piece of sandpaper to the front edge of the block of wood, and bend the sandpaper underneath the block so that the sandpaper will be in contact with the board.
6. Pull the block with the sandpaper underneath it across the board. Does this change in the nature of the surfaces in contact affect the force of friction?
7. Remove the sandpaper and put some dowel sticks or round pencils under the block of wood. How does this affect the force of friction?

Going Further: Try to find examples, such as brakes and walking, where friction is needed. What would it be like to live in a world without friction? Try to find ways in which we try to increase the force of friction. Try to find as many examples as possible where friction is reduced.

8

Exploring the Universe: Sun, Moon, and Stars

INTRODUCTION

It is important that children have firsthand experiences observing sun, moon, planets, stars, and other astronomical objects. Some children only study the universe through books, newspapers, and occasional television programs. In part because a number of these observations have to be made at night, there is a danger that these children will not gain a sense of the reality of astronomical objects. Unfortunately, this is especially the case in cities, where it can be difficult to see some of the objects and patterns in the nighttime sky. Every child should have the experience of studying the sun, moon, stars, and planets directly.

"It's great!" "I never knew there were so many stars!" These are the kinds of exclamations you will hear when children have a chance to see the nighttime sky in all its grandeur. As parents, teachers, and friends, we can take children out to see the nighttime sky individually, on "star parties," or as part of a camping experience, or we can guide them to make the observations on their own. Some observations can be made almost anywhere, but the stars are at their brightest away from the bright lights of the cities.

Observations of the sun may be suggested. However, children should be cautioned against looking directly at the bright sun. Activities suggested in this chapter include the study of shadows and the observation of the sun near sunrise and sunset when the sun is not as bright.

SUNRISE AND SUNSET

Investigation: Where does the sun rise? Where does the sun set? (P, I)

Background Information: The places where the sun seems to rise and set change dramatically with the seasons. Actually, this is because the earth revolves around the sun once a year and because the earth's axis is tilted. Therefore, to people in the Northern Hemisphere, the sun appears lower in the southern sky in the winter than it does in the summer. Children can check and record the places where the sun seems to rise and set during each of the seasons.

Materials: Pencil
Paper

Procedure: 1. Have the children make sketches of the eastern and western horizons. Make certain that they include in their sketches prominent objects on the horizon such as trees, hills, and tall buildings.
2. Have them observe a sunrise and a sunset and note the points on the horizon where the sun seems to rise and set.
3. Ask the children to draw on the horizons they have sketched the points where the sun seems to rise and set. Ask them to draw these points relative to some prominent object on the horizon (see Figure 8-1).
4. Date the sketches.
5. Repeat the observations each season. The greatest changes in relative position will be noted if some observations are made near the winter solstice (December 22) and the summer solstice (June 21).

Going Further: The path that the sun seems to take across the sky also seems to change with the seasons. One way to check on this is to note how far the sun's light extends at noon into a room having a southern exposure. If possible, put a small piece of tape on the floor at the tip of the sunlight. Then do the same at each season. How does the extent to which sunlight shines into the room change with the seasons?

Similar observations and records can be made and kept of where certain stars in the east appear early in the evening. Do their positions seem to change with the seasons?

June 1

FIGURE 8-1. An eight-year-old's sketch of where the sun seems to rise on the eastern horizon. Note how close the sun seems to be to the factory.

CHANGING SHADOWS

Investigation: How do the shadows change during the day? (P, I)

Background Information: Because of the rotation of the earth, the sun appears to move across the sky from east to west during the course of a day. This apparent movement of the sun is so regular that it can be used to indicate time. However, since the sun is so bright that we should not look directly at it, a good way to study this apparent movement of the sun across the sky is to study the changes in shadows.

Materials: Sharpened pencil
Spring clothespin or small piece of clay
Piece of paper
Pencil
A flat surface that can be in the sunlight during the course of the day

Procedure:
1. Find a flat surface that is in the sunlight the entire day. This flat surface can be just inside a window on the south side of a room, or somewhere outside.
2. Place the blunt end of a sharpened pencil between the two prongs of a spring clothespin or into a small piece of clay so that the pencil is supported in a vertical position. If a spring clothespin is used, it should be taped to the surface so that it will not tip.
3. Place a sheet of paper on the flat surface to the north of the upright pencil so that the shadows can be marked on the paper. Fasten the paper with pieces of tape so that it will not move (see Figure 8-2).
4. Beginning early in the morning, place a dot at the tip of the shadow each hour during the course of the day. How does the direction of the shadow change during the day? How does the length of the shadow change during the day?

Going Further: Repeat the investigation at a different season of the year. How do the paths and lengths of shadows change with the seasons?
With the help of a clock or watch, mark the position of the shadow at each hour so that the procedure can be used to indicate time.

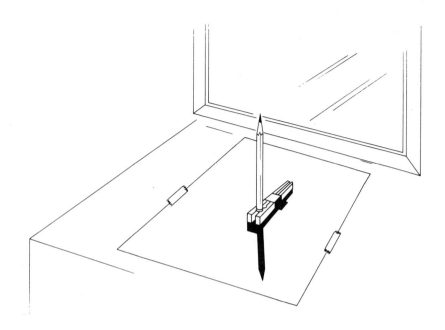

FIGURE 8-2. Mark the location of the tip of the pencil's shadow each hour.

PATTERNS IN THE SKY

Investigation: How can various constellations be located?   (I)

Background Information: The stars in the sky seem to stay in the same position relative to each other. They seem to form patterns, which we call *constellations.* The ancients named the constellations after animals or well-known objects. To see and identify some of these constellations can be a real thrill for children. A good place to start is with the stars in the northern sky.

Materials: Good pair of eyes
Star maps such as the one in Figure 8-3.

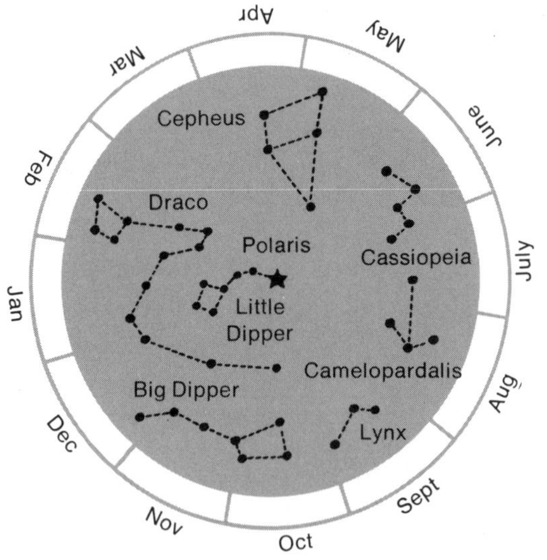

FIGURE 8-3.   A map of some of the stars in the northern sky.

Procedure: 1. Figure 8-3 is a map of the constellations around *Polaris*, the North Star. Turn the map so that the current month is to the north. Then the sky map shows the constellations as they appear at 9 P.M. *standard time.* For every hour after 9 P.M., turn the map a half month in a counterclockwise direction.

2. *North Star.* Locate the Big Dipper. Stars A and B of the Big Dipper are called the pointers. Draw an imaginary line through A and B and extend it about five times the distance between A and B. The star at the end of this imaginary line is Polaris, the North Star.

3. *Little Dipper.* Polaris is at the end of the handle of the Little Dipper.

4. *Cepheus.* It resembles a church steeple and is on the opposite side of the North Star from the Big Dipper.

5. *Cassiopeia.* Draw an imaginary line from the middle of the Big Dipper to the North Star and extend it an equal distance on the other side. The chairlike constellation is Cassiopeia.

Going Further: With the help of star maps, children can locate constellations in other parts of the sky. Current star maps are available in such publications as:

Science and Children. Washington, D.C.: National Science Teachers Association.

Natural History. New York: American Museum of National History.

Sky and Telescope. Cambridge, Mass.: Sky Publishing Corporation.

If there is a planetarium in your locality, they will usually have maps, charts, and other materials available.

PHASES OF THE MOON

Investigation: How does the shape of the moon seem to change?   (I)

Background Information: We can see the moon because it is illuminated by light from the sun. The moon revolves around the earth once every month, and that fraction of the illuminated part of the moon that we can see changes throughout the month.

At *new moon*, the moon is between the sun and the earth, and we cannot see it at all. Just after the new moon, we will see the thin sliver of the *waxing moon* low in the western sky just after sunset. (The appearance of this thin silver of the waxing moon at times signals the beginning of religious observances in some cultures.) Each evening after that, it will be possible to see an increasing amount of the illuminated part of the moon. At *full moon*, the earth is between the sun and the moon, and the moon appears to be a bright disk in the sky. The moon continues to rise about 50 minutes later, and each evening we will see a smaller *waning moon*. As the month progresses, we can see the waning moon in the daytime. When the moon has again revolved to a position between the sun and the earth, a month has passed and it is again new moon.

Some children may ask about eclipses. A *solar eclipse* occurs when the shadow of the moon falls on some part of the earth. A *lunar eclipse* occurs when the shadow of the earth falls on the moon. However, we do not have eclipses each month because the path of the moon's revolution around the earth is at a slight angle with the plane of the earth's path around the sun.

Materials: Pencil

               Paper

Procedure: 1. Observe the moon. Many newspapers and calendars carry information on the phases of the moon and the time of moonrise. Make a sketch of the shape of the moon. Note the position of the moon in the sky. Write the date and time of the observation under the sketch.

            2. Try to observe and make sketches of the shape of the moon each day. Note the time of the observation and the position of the moon in the sky (see Figure 8-4). Of course, it will be impossible to make these observations on cloudy days and nights, but try to make as many observations as possible over the course of a month.

            3. Question the children about the observations of apparent changes in the shape of the moon. Explanations of the phases of the moon are available in many books, including Jacobson and Bergman, *Science for Children* (Englewood Cliffs, N.J.: Prentice-Hall, Inc., 1980), pp. 306-308.)

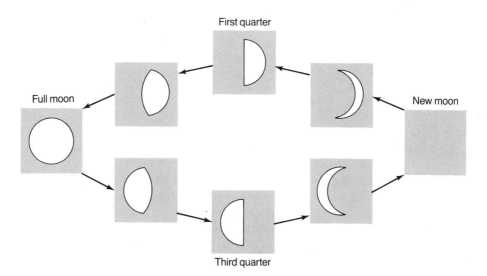

FIGURE 8-4. The shape of the moon seems to change during the month.

Going Further: Look for "earthshine." The earth is a good reflector of sunlight, and some of this sunlight is reflected to the moon and then back to the earth. At times, we can see parts of the moon that are dimly illuminated with "earth-shine."

Observe the moon with binoculars or a telescope. Try to identify such features as craters, the dark patches (called *maria*), and the light-colored rays.

When lunar or solar eclipses occur, make every effort to give the children an opportunity to see them.

ROCKETS TO EXPLORE SPACE

Investigation: How does a rocket work?   (P, I)

Background Information: A rocket is an engine that can travel in space. It carries within it everything that is needed to propel it. Perhaps the simplest rocket is an inflated balloon.

When a balloon is inflated and then released, it will move through space. We can think of the air in a balloon as pushing against all sides of the balloon. You can feel this as you press your fingers against the balloon. The balloon does not move because the pushes against one side of the balloon are balanced by the pushes against the opposite side. But when the air in the balloon is released, the push against the front of the balloon is not balanced by a push against the opening in the balloon, and the balloon moves.

Materials: String (A fish line works very well.)
Balloon
Drinking straw
Masking tape or Scotch tape

Procedure: 1. Cut a drinking straw in half and insert the string through one half of the drinking straw.
2. Fasten one end of the string to a doorknob or some other support at one side of the room. Extend the string to the other side of the room. Pull it taut and fasten it to another support. Test to see that the drinking straw slides smoothly on the string.
3. Tear off two pieces of tape each about 10 centimeters (4 inches) long.
4. Inflate the balloon and hold it closed with the thumb and forefinger of one hand. Fasten an end of each of the pieces of tape to the top side of the balloon. Stretch the pieces of tape across the drinking straw, and attach the second end of each piece of tape to the balloon. Press the tapes down onto the drinking straw so that the tapes will stick to the drinking straw (see Figure 8-5).

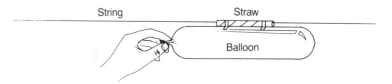

FIGURE 8-5.   A balloon rocket made by taping a balloon to a drinking straw that slides on a string.

5. Slide the balloon and the attached drinking straw to one end of the string.

6. Release the balloon so that the air inside the balloon can escape. What happens? How far does the balloon slide along the string?

7. Try to invent ways to make the balloon slide farther and faster. For example, insert a straw into the balloon opening and hold the straw in place in the balloon opening with a rubber band. Now, when the balloon is released, the air will pass through the straw.

Going Further: Various kinds of rockets are available as toys. One kind, for example, ejects water under pressure and can be propelled quite high into the air. Some of these rockets even have more than one stage. Try to obtain one or more of these rockets. Be sure to read the directions carefully. Most of them should be set off outdoors.

When possible, view the launch of spacecraft on television. What is emitted by these rockets to propel the spacecraft?

TIME

Investigation: How can we measure time? (P, I)

Background Information: Time has been called the fourth dimension. Time is usually measured with some phenomenon that is periodic. Measurement of the time of day is often based upon the rotation of the earth, and the time of year is based upon the revolution of the earth around the sun. In this activity, a seconds pendulum is constructed that can be used to measure time in terms of the swing of the pendulum. A seconds pendulum is about 99 centimeters (39 inches) long.

Materials: Small weight
String
Watch or clock with second hand
Meter stick

Procedure: 1. Ask the children to suggest various ways that we keep time. If there is a clock with a pendulum nearby, have them watch the swing and note how regular the swing is. Suggest that they try to make a pendulum in which the weight will complete a swing in one direction in one second.
2. Attach a small weight to the end of a string. Hang it so that the weight is free to swing and so that the length of the pendulum can be adjusted.

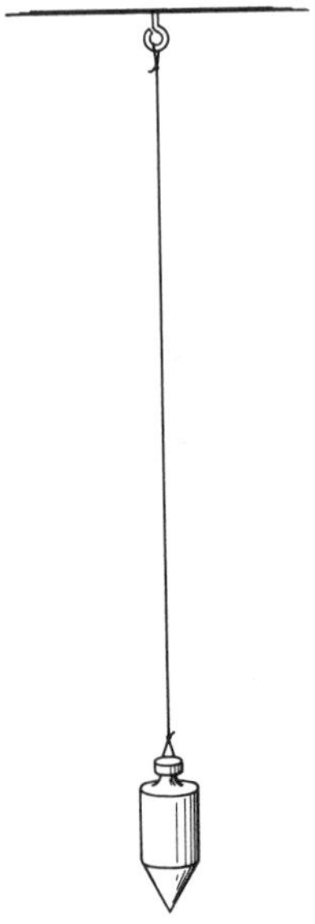

FIGURE 8-6. Construct a "seconds" pendulum that will swing 10 swings in either direction in 10 seconds.

3. Have the children pull the weight back, release it, and count the number of swings in either direction in 10 seconds. (With very young children, you may have to keep time; with others, this may be an ideal time to teach them how to use a watch or clock.)

4. Adjust the length of the pendulum until it makes 10 swings in either direction in 10 seconds. If a pendulum is making more than 10 swings in 10 seconds, it should be lengthened; if fewer than 10 swings, it should be shortened.

5. Have the children measure the distance from the center of the weight to where the string is attached.

Going Further: Children can count the number of swings of a pendulum in a pendulum clock. Are these seconds pendulums?

Children can measure time using their pulse. How many times does their pulse beat in 10 seconds?

PLANETS: THE WANDERERS

Investigation: How do positions of planets change? (P, I)

Background Information: In addition to Earth, five planets are easily visible to the unaided eye. They are Mercury, Venus, Mars, Jupiter, and Saturn. These planets can usually be seen very near the imaginary path of the sun across the sky. Venus and Mercury, since they are planets nearest the sun, will always be seen near the sun. Sometimes one or both of the planets can be seen in the western sky just after sunset. Venus is so bright that it sometimes can be seen in daylight, if we know just where to look. Or one or both can sometimes be seen in the eastern sky just before sunrise. The other visible planets can be seen at various times near the imaginary path of the sun.

We can see the planets because they reflect sunlight to us. Usually the sunlight reflected by the planets shines steadily and does not twinkle like starlight.

Although the stars always remain in the same positions relative to each other, the planets appear to move over a period of days and weeks relative to the background stars. The word *planet* is said to mean "wanderer." In this activity, some of this wandering will be charted.

Information on identification and location of planets can be obtained from such journals as:

Science and Children. Washington, D.C.: National Science Teachers Association.

Natural History. New York: American Museum of Natural History.

Sky and Telescope. Cambridge, Mass.: Sky Publishing Corporation.

Many daily newspapers and world almanacs publish the times for the rising and setting of the planets. (In some sources the planets are erroneously referred to as evening or morning stars.)

Materials: Pencil

Paper

Procedure:
1. Locate the planets that are visible in the sky. The visible planets are usually quite bright, are found along the path that the sun follows in the sky, and usually do not twinkle. (The publications listed above will be helpful in locating planets.)
2. Note the positions of the planets relative to a prominent nearby star. One of the ways to determine the position is to hold a clenched fist at arm's length and estimate the number of fists the planet is from the prominent star.
3. Mark the relative position of the planet and the star on a sheet of paper. Use some convenient scale, such as one centimeter equals one fist. Mark the date near the position of the planet.

4. Every three or four days note and mark the relative positions of the planet and star. Does the planet seem to move relative to the star? If so, how?
5. Note the relative positions of other planets and nearby prominent stars. Do these planets seem to move?

Going Further: Try to locate the planet Mercury. Mercury is always near the sun and can only be seen just after the sunset or before sunrise.

Research some of the mythology associated with the planets. For example, how did each of the planets get its name? What is the significance of the name?

DISTANCES IN SPACE

Investigation: How far away is it?   (I)

Background Information: It is important in astronomy to try to find out how far away stars, planets, and other celestial objects are. We have to find indirect ways to measure these distances, and we can use similar indirect ways to measure the distances across a lake or river or the distances across a backyard or school grounds.

Materials   Protractor
Meter stick
Drinking straw
Pin

Procedure: 1. Tape a protractor to a meter stick so that the base of the protractor is parallel with the edge of the meter stick.
2. Stick a pin through the drinking straw and the hole in the center of the protractor. The pin should be one-third of the distance up the drinking straw, and it should be easy to turn the drinking straw around the pin (see Figure 8-7).

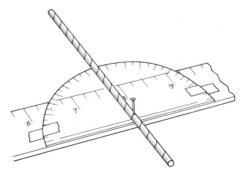

FIGURE 8-7.   Use a drinking straw and a protractor to measure angles.

3. Place the protractor and meter stick on a support, such as the back of a chair. Move the drinking straw so that it is directly over the 90 degree mark. Look through the drinking straw and line it up so that you can see the object (Point X) whose distance away you wish to find.
4. Place the meter stick on the floor or ground. Using a taut string, draw a straight line as an extension of the edge of the meter stick. This line should be a known length, such as 10 meters.

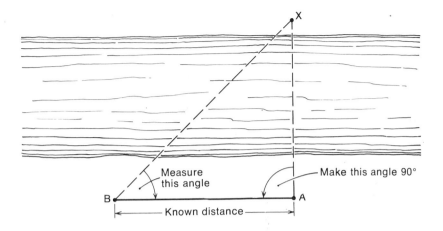

FIGURE 8-8. Measure the angle at B. Then make a scale drawing of the triangle and measure the distance from A to X on the drawing.

5. Move the meter stick and protractor to the other end of this line (Point B in Figure 8-8).
6. Again, put the protractor and meter stick on a support so that the meter stick and the protractor are parallel with the base line that you have drawn.
7. Sight through the drinking straw and turn it until the distant point (Point X) can be seen through the drinking straw. Note the angle on the protractor between the base line and the drinking straw.
8. Make a scale drawing. Use a convenient scale, such as 1 centimeter equals 1 meter. With such a scale, draw a base line (10 centimeters). Draw a line at right angles (90 degrees) at one end of the base line (Point A). At the other end of the base line (Point B), draw a line at the same angle as the angle between the drinking straw and the base of the protractor. Extend this line on the scale drawing until it crosses the line drawn at 90 degrees from the base line at point A. Measure the distance on the scale drawing from point A to point X. Multiply this distance by the scale of the drawing. For example, if the scale is 1 centimeter equals 1 meter, multiply the distance from A to X on the drawing by 100. This will give you the actual distance from Point A to Point X.

Going Further: Astronomers use a method similar to this to measure the distance to planets and nearby stars. Children can use this method to measure the heights of buildings, flagpoles, and other tall objects.

Children can also use mathematic skills to compare the length of the shadow of a building of unknown height to the shadow of an object of known height.

$$\frac{\text{Unknown height}}{\text{Length of its shadow}} = \frac{\text{Height of object of known height}}{\text{Length of its shadow}}$$

$$\text{Unknown height} = \frac{\text{Height of known object} \times \text{length of shadow of object of unknown height}}{\text{Length of shadow of known height}}$$